Lecture Notes in Control and Information Sciences

Edited by M. Thoma and A. Wyner

Lecture Notes in Control and Information Sciences

Edited by M. Thoma and A. Wyner

150

L.C.G.J.M. Habets

Robust Stabilization in the Gap-topology

Springer-Verlag
Berlin Heidelberg GmbH

Author

Luc C. G. J. M. Habets
Eindhoven University of Technology
Dept. of Mathematics and Computing Science
Dommelbuilding 1.11
P.O. Box 5 13
5600 MB Eindhoven
The Netherlands

ISBN 978-3-540-53466-2   ISBN 978-3-540-46889-9 (eBook)
DOI 10.1007/978-3-540-46889-9

61/3020-543210   Printed on acid-free paper.

Crescit scribendo scribendi studium

(Desiderius Erasmus)

PREFACE

'Crescit scribendo scribendi studium'. This experience I had, while writing this book. When I started, I thought I had everything clearly in mind, but, while writing, things turned out to be more complicated. The obligation to state results precisely, forced me to treat the subject in more detail than I planned. I hope however that a greater insight of the reader in the various notions and subtility of the theory will be the gain of this effort. But it took me more paper to tell the story than I expected, and I hope that this final result is clarifying instead of tedious or boring to the reader.

Roughly speaking, this book can be divided into three main parts. In the first one, the notions of robust stabilization and the gap–topology are introduced. In the second part, most of the theory is developed. We find sufficient conditions for robust stabilization in the gap–topology, and are able to derive a method for the solution of the problem of optimally robust control. Finally, in the third part, algorithms are developed to actually compute a solution to this problem for systems in state–space form. So in this book we slowly tend from theory to practice.

Of course I'm grateful to all the people who helped me, in one way or the other, with the accomplishment of this paper. First of all I'd like to thank prof. M.L.J. Hautus. He awoke my interest for the field of systems theory and also attended me to this subject. I am also indebted to Siquan Zhu, who developed most of the theory, upon which my algorithms are based. Especially I would like to thank dr. Kees Praagman for carefully reading the manuscript and all his valuable suggestions. The cooperation with him was very stimulating and I'd like to thank him for always being available when I had questions or problems. I'm also grateful to drs. Anton Stoorvogel. When I didn't understand the computer anymore, he always got me out of trouble. Finally I'd like to mention drs. J. Timmermans. Without him, this book would probably have never been written. And last but not least, I thank my parents for all their support and encouragements. Without them I'd never come so far.

Luc Habets.

CONTENTS

NOTATION

Given a state–space realization [A,B,C,D] of a plant P, the transfermatrix corresponding to P is given by

$$P(s) = D + C(sI-A)^{-1}B.$$

Throughout the text we use the notation [A,B,C,D] not only to describe the realization of P, but also to indicate the related transfermatrix, i.e.

$$[A,B,C,D] = D + C(sI-A)^{-1}B.$$

Using this data–structure we can easily obtain the following formulae for operations on transfermatrices.

$$[A,B,C,D] = [T^{-1}AT, T^{-1}B, CT, D].$$

$$[A,B,C,D]^{-1} = [A-BD^{-1}C, BD^{-1}, -D^{-1}C, D^{-1}].$$

$$[A_1,B_1,C_1,D_1] + [A_2,B_2,C_2,D_2] =$$
$$= \left[\begin{bmatrix} A_1 & 0 \\ 0 & A_2 \end{bmatrix}, \begin{bmatrix} B_1 \\ B_2 \end{bmatrix}, (C_1,C_2), D_1+D_2 \right].$$

$$[A_1,B_1,C_1,D_1] * [A_2,B_2,C_2,D_2] =$$
$$= \left[\begin{bmatrix} A_1 & B_1C_2 \\ 0 & A_2 \end{bmatrix}, \begin{bmatrix} B_1D_2 \\ B_2 \end{bmatrix}, (C_1,D_1C_2), D_1D_2 \right]$$
$$= \left[\begin{bmatrix} A_2 & 0 \\ B_1C_2 & A_1 \end{bmatrix}, \begin{bmatrix} B_2 \\ B_1D_2 \end{bmatrix}, (D_1C_2,C_1), D_1D_2 \right].$$

0. INTRODUCTION

This book addresses the problem of robust stabilization. Given a plant P we often encounter the problem of stabilization: find a compensator C such that the transferfunction of the closed–loop–system is stable. But when P is stabilizable, there is often a whole set of stabilizing compensators. Now we get the problem which one to choose. So we need a criterion to select a certain stabilizing compensator C.

In practice however, the description of the plant P is often inaccurate (for example because the plant–description is approximated by statistical means). So we don't work with the real plant P, but with a model of it, say P_m. Now, when we construct a compensator C, that stabilizes P_m, the question is whether it stabilizes P also. When P and P_m differ only slightly, we expect C to do so. But how much P and P_m may differ?

Now we can describe the problem of robust stabilization. Given a plant P, we want to find a compensator C, that not only stabilizes P, but also a neighborhood of P. By changing C, we will also change the stabilized neighborhood. Now we are interested in that compensator C, that stabilizes the largest neighborhood of P.

In practice this means the following. Suppose we have a (probably inaccurate) model P_m of the plant P. We stabilize P_m by that compensator C, that also stabilizes the greatest neighborhood of P_m. So, when the real plant P is in that neighborhood, it is also stabilized by C. In this way we have designed a robust compensator C.

Now we will give a short outline of the contents of this paper. In the first chapter we will make the above statements more precise. We give a definition of stability and parametrize all stabilizing compensators. At the end we give a more formal description of the problem of robust stabilization.

In the second chapter we introduce the gap–metric. This metric gives a measure for the distance between two plants. The gap–metric, and the gap–topology, induced by this metric, will turn out to be very important for the robust stabilization problem, because they determine whether two plants are close to each other or not.

With the help of the gap–metric, we are able to derive sufficient conditions for robust stabilization. This is the content of chapter 3.

Using the bounds for robust stabilization found in chapter 3, we will tackle the problem of optimally robust control in chapter 4. We derive a method that gives an optimally robust compensator for each plant P.

In chapter 5 we translate the theoretical method of chapter 4 to an algorithm, that actually computes a solution of the problem, for systems in state–space form.

Unfortunately the algorithm, described in chapter 5, gives solutions of a very high order. This means that in practice the solution is hardly applicable. In chapter 6 we describe an algorithm that gives solutions of a much lower order. Also an other method will be discussed, which decreases the order of the solution drastically.

Finally in chapter 7 we draw our conclusions and give a short outline for eventual future developments in this field.

At the end of this introduction we remark that, certainly in the first part of this text, we follow along the same lines as Zhu in [19]. In this dissertation however, Zhu treats the same problem in a more general context. We confine ourselves to the case that all the entries of the transfermatrix of a plant are *rational* functions. We often even assume these functions to be *real–rational* (this case represents the continuous–time lumped LTI–systems).

1. ROBUST STABILIZATION

In this chapter we first give definitions of stability and stabilizability. Then we introduce the concept of Bezout factorizations of a plant P. With the help of these factorizations we are able to give conditions for the stabilizabity of a plant P, and parametrize all stabilizing compensators. Finally we give a more precise characterization of the problem of robust stabilization.

1.1. Stability and stabilizability

Let $\mathbb{R}(s)$ denote the set of all rational functions in s with real coefficients, and $M(\mathbb{R}(s))$ the set of all matrices with elements in $\mathbb{R}(s)$. So $M(\mathbb{R}(s))$ consists of all matrices with real–rational functions as entries. Throughout this text we usually assume that the transferfunction $P(s)$ of a plant P is an element of $M(\mathbb{R}(s))$.

By $\mathbb{R}_p(s)$ we denote the set of all *proper* rational functions in s with real coefficients. So $\mathbb{R}_p(s)$ is a subset of $\mathbb{R}(s)$ consisting of all rational functions in $\mathbb{R}(s)$, for which the degree of the numerator is not larger than the degree of the denominator. Again $M(\mathbb{R}_p(s))$ consists of all matrices with elements in $\mathbb{R}_p(s)$. In the last chapters of this paper we will assume that the transfermatrix $P(s)$ of a plant P belongs to $M(\mathbb{R}_p(s))$.

Now we introduce the space H_∞, which plays an important role in the problem of robust stabilization.

DEF.1.1.1. The *space* H_∞ consists of all functions $F(s)$ of the complex variable s, which are analytic in Re s > 0, take values in $\mathbb{C}$, and are bounded in Re s > 0, i.e.

$$\sup \{ \ | \ F(s) \ | \ | \ \text{Re s} > 0 \ \} < \infty \tag{1.1}$$

The left–hand side of (1.1) defines the H_∞–*norm* of a function $F \in H_\infty$. With this norm, H_∞ is a Banach–space.

It is a well known fact that we can replace the open right half–plane in (1.1) by the imaginary axis. This gives the following, mostly used formula for the norm of a function $F \in H_\infty$:

$$\| \ F \ \|_\infty = \sup \{ \ | \ F(i\omega) \ | \ | \ \omega \in \mathbb{R} \ \} \tag{1.2}$$

DEF.1.1.2. $\mathbb{R}H_\infty := \mathbb{R}(s) \cap H_\infty$

Note that $\mathbb{R}H_\infty$ is a subspace of H_∞. We also remark that $\mathbb{R}H_\infty$ is a subspace of $\mathbb{R}_p(s)$, because a real rational function in H_∞ has to be proper. Let, analogous as before, $M(H_\infty)$ and

M(RH$_\infty$) denote the sets of all matrices with elements in H$_\infty$ and RH$_\infty$ respectively. Then we can now give a definition of stability:

DEF.1.1.3 A plant *P* is *stable* when its transferfunction P(s) is an element of M(H$_\infty$).

Under the assumption P(s) $\in$ M(R(s)), (1.1.3) becomes: a plant *P* with transferfunction P(s) $\in$ M(R(s)) is stable if P(s) $\in$ M(RH$_\infty$). Remark that the notion of stability we have introduced here, is BIBO–stability (bounded–input, bounded–output). A real–rational transferfunction is stable if it is proper, and all its poles have negative real parts.

We now introduce the *feedback–system* shown in figure (1.3), where *P* represents the plant (with transferfunction P), and *C* a compensator (analogous to the plant *P*, we assume that the entries of the transfermatrix C of the compensator *C* are all real–rational functions, so C $\in$ M(R(s))).

(fig.1.3)

u_1, u_2 denote *external inputs*, e_1, e_2 *inputs* to the compensator and the system respectively, and y_1 and y_2 *outputs* of the compensator and the plant respectively. This model is versatile enough to accommodate several control problems. Later on we will refer to this set–up as a feedback–system.

Suppose P,C $\in$ M(R(s)). The transfermatrix from $u := \begin{bmatrix} u_1 \\ u_2 \end{bmatrix}$ to $e := \begin{bmatrix} e_1 \\ e_2 \end{bmatrix}$ is given by (see [15, p.101]):

$$H(P,C) := \begin{bmatrix} (I+PC)^{-1} & -P(I+CP)^{-1} \\ C(I+PC)^{-1} & (I+CP)^{-1} \end{bmatrix} \qquad (1.4)$$

where we assumed that P and C have compatible dimensions, and the well–posedness condition $|I+CP| = |I+PC| \neq 0$, is also satisfied; so H(P,C) makes sense. In this case, the transferfunction W(P,C) from u to $y := \begin{bmatrix} y_1 \\ y_2 \end{bmatrix}$ is given by:

$$W(P,C) = \begin{bmatrix} 0 & I \\ -I & 0 \end{bmatrix}(H(P,C)-I) = \begin{bmatrix} C(I+PC)^{-1} & -CP(I+CP)^{-1} \\ PC(I+PC)^{-1} & P(I+CP)^{-1} \end{bmatrix} \qquad (1.5)$$

We can now give the following definition of stabilizability:

DEF.1.1.4. A system $P \in M(\mathbb{R}(s))$ is said to be *stabilizable* if there exists a compensator $C \in M(\mathbb{R}(s))$ such that $|I+PC| \neq 0$, and the transferfunction W(P,C) from u to y is stable. If W(P,C) is stable, C is called a *stabilizing compensator of P*.

From formula (1.5) we see immediately that W(P,C) is stable if and only if H(P,C) is stable. Because H(P,C) has a simpler form than W(P,C), we mostly use the following characterization of stabilizability (the proof is obvious from the argument above).

THEOREM 1.1.1. A system $P \in M(\mathbb{R}(s))$ is stabilizable iff there exists a compensator $C \in M(\mathbb{R}(s))$ such that $|I+PC| \neq 0$ and the transferfunction H(P,C) from u to e is stable.

Finally we remark that the conditions for stability are symmetric in P and C, i.e. H(P,C) is stable iff H(C,P) is stable. This because

$$\begin{bmatrix} 0 & -I \\ I & 0 \end{bmatrix} H(P,C) \begin{bmatrix} 0 & I \\ -I & 0 \end{bmatrix} = H(C,P) \tag{1.6}$$

1.2. Bezout factorizations; parametrization of all stabilizing controllers

In this section we introduce the concept of right– and left–Bezout factorizations. With help of these factorizations we can give conditions for the stabilizability of a plant P. Finally we are able to parametrize all compensators C, that stabilize a certain plant P.

DEF.1.2.1. Let P an n×m transfermatrix, $P \in M(\mathbb{R}(s))$. We say $(D,N) \in M(\mathbb{R}H_\infty)$ is a *right–Bezout factorization* (r.b.f.) of P if:
    1) D is an m×m, and N an n×m transfermatrix, and $|D| = \det(D) \neq 0$;
    2) There exist two matrices Y and Z in $M(\mathbb{R}H_\infty)$ such that
$$YD + ZN = I \tag{1.7}$$
    3) $P = ND^{-1}$

Similarly we can define left–Bezout factorizations:

DEF.1.2.2. Let P an n×m transfermatrix, $P \in M(\mathbb{R}(s))$. We say $(\tilde{N},\tilde{D}) \in M(\mathbb{R}H_\infty)$ is a *left–Bezout factorization* (l.b.f.) of P if:
    1) $\tilde{D}$ is an n×n, and $\tilde{N}$ an n×m transfermatrix, and $|\tilde{D}| = \det(\tilde{D}) \neq 0$;
    2) There exist two matrices $\tilde{Y}$ and $\tilde{Z}$ in $M(\mathbb{R}H_\infty)$ such that
$$\tilde{D}\tilde{Y} + \tilde{N}\tilde{Z} = I \tag{1.8}$$
    3) $P = \tilde{D}^{-1}\tilde{N}$

Remark that an r.b.f. (resp. l.b.f.) of a $P \in M(\mathbb{R}(s))$ is unique up to right (resp. left) multiplication by unimodular matrices.

In our case $(P \in M(\mathbb{R}(s)))$ the concept of Bezout factorizations is completely analogous to the notion of coprime factorizations introduced by Vidyasagar in [15, pp.74–75]. Since in our case there is no difference between these two ideas, we can use both these terms to point out the same concept.

We can now ask the question: when does a transfermatrix $P \in M(\mathbb{R}(s))$ have a Bezout factorization? The following theorem gives the answer.

THEOREM 1.2.1. Each $P \in M(\mathbb{R}(s))$ has a right– and a left–Bezout factorization.

PROOF The proof of this theorem can be found in [15, p.75]. A constructive proof for the case $P \in M(\mathbb{R}_p(s))$ is given in [4, pp.23–25]. Later on, in chapter 4, we will give a constructive proof of the existence of so called normalized Bezout factorizations for this case. These are a special sort Bezout factorizations. So for $P \in M(\mathbb{R}_p(s))$ that proof is also sufficient for the existence of ordinary right– and left–Bezout factorizations.

For the sake of completeness we mention the fact that for other (more general) classes of transfermatrices, the existence of right– or left–Bezout fractions is not always guaranteed (for example $f(s) = se^{-s}$ doesn't have a right– or left–Bezout factorization in H_∞).

With the help of the concept of Bezout factorizations we can now give a condition for the stabilizability of a plant.

THEOREM 1.2.2. If the transfermatrix P of a plant P has a right–Bezout factorization, then P is stabilizable. Moreover, any stabilizing controller C of P, has a left–Bezout factorization.

PROOF First we show that if P has an r.b.f. then P is stabilizable. Assume (N,D) is an r.b.f. of P. Now suppose there are matrices Y and Z such that

$$YD + ZN = U$$

is invertible. Suppose $|Y| \neq 0$. When we now define $C := Y^{-1}Z$, then

$$H(P,C) = \begin{bmatrix} I & 0 \\ 0 & 0 \end{bmatrix} + \begin{bmatrix} -N \\ D \end{bmatrix} (YD + ZN)^{-1} (Z,Y). \qquad (1.9)$$

We prove (1.9) by checking the four equalities of formula (1.4).
1) $(I+PC)^{-1} = I - N(YD+ZN)^{-1}Z.$
We show $(I+PC)(I-NU^{-1}Z) = I.$
$(I+PC)(I-NU^{-1}Z) = (I+ND^{-1}Y^{-1}Z)(I-NU^{-1}Z) =$

$= I-NU^{-1}Z+ND^{-1}Y^{-1}Z-ND^{-1}Y^{-1}ZNU^{-1}Z = I - N(U^{-1}-D^{-1}Y^{-1}+D^{-1}Y^{-1}ZNU^{-1})Z =$

$= I-N(U^{-1}-D^{-1}Y^{-1}+D^{-1}Y^{-1}(U-YD)U^{-1})Z = I - N(U^{-1}-D^{-1}Y^{-1}-U^{-1})Z = I.$

2) $C(I+PC)^{-1} = D(YD+ZN)^{-1}Z.$

$C(I+PC)^{-1} = Y^{-1}Z(I-NU^{-1}Z) = Y^{-1}Z - Y^{-1}ZNU^{-1}Z = Y^{-1}Z - Y^{-1}(U-YD)U^{-1}Z =$

$= Y^{-1}Z - Y^{-1}Z + DU^{-1}Z = D(YD+ZN)^{-1}Z.$

3) $(I+CP)^{-1} = D(YD+ZN)^{-1}Y.$

We show $(I+CP)DU^{-1}Y = I.$

$(I+CP)DU^{-1}Y = (I+Y^{-1}ZND^{-1})DU^{-1}Y = DU^{-1}Y + Y^{-1}ZNU^{-1}Y =$

$= DU^{-1}Y + Y^{-1}(U-YD)U^{-1}Y = DU^{-1}Y + I - DU^{-1}Y = I.$

4) $-P(I+CP)^{-1} = -N(YD+ZN)^{-1}.$

$-P(I+CP)^{-1} = -ND^{-1}D(YD+ZN)^{-1}Y = -N(YD+ZN)^{-1}Y.$

Now we know that (N,D) is an r.b.f. of P. So there exist matrices Y and Z in $M(\mathbb{R}H_\infty)$ such that

$$YD + ZN = I.$$

When $|Y| \neq 0$, we see immediately that when we choose $C = Y^{-1}Z$, then

$$H(P,C) = \begin{bmatrix} I & 0 \\ 0 & 0 \end{bmatrix} + \begin{bmatrix} -N \\ D \end{bmatrix}(Z,Y).$$

Because N,D,Y and Z are all matrices in $M(\mathbb{R}H_\infty)$, $H(P,C) \in M(\mathbb{R}H_\infty)$ and so C stabilizes P.

Now suppose $|Y| = 0$. Let K be a matrix in $M(\mathbb{R}H_\infty)$ of the same size as Y and such that $(Y^T, K^T)^T$ has full column rank. Define

$$V := \{ R \in M(\mathbb{R}H_\infty) \mid |Y + RK| \neq 0 \}.$$

Then V is an open and dense subset of the set of all matrices in $M(\mathbb{R}H_\infty)$ of the same size as Y (for a proof, see [15, p. 111]). Now we take an $R \in V$ such that $\| RKD \| < 1$. Then $(I + RKD)$ is invertible, with bounded inverse (for a proof, see lemma 3.2.2.) and because of the Bezout–identity (1.7) we have

$$(Y+RK)D + ZN = RKD + YD + ZN = I + RKD.$$

Since $(I + RKD)$ is invertible and $|Y + RK| \neq 0$, we have, when we choose $C := (Y+RK)^{-1}Z$,

$$H(P,C) = \begin{bmatrix} I & 0 \\ 0 & 0 \end{bmatrix} + \begin{bmatrix} -N \\ D \end{bmatrix}(I+RKD)^{-1}(Z,Y+RK).$$

Because all the matrices on the right–hand side belong to $M(\mathbb{R}H_\infty)$, $H(P,C) \in M(\mathbb{R}H_\infty)$, and C stabilizes P.

The second part of the proof, the fact that each stabilizing controller has an l.b.f., is quite clear from the exposition above. For an exact proof, however, we refer to [15, p.363].
□

Analogous to theorem 1.2.2., the following holds: if P has a left–Bezout factorization, then P is stabilizable, and any stabilizing compensator C has a right–Bezout factorization. From theorems 1.2.1. and 1.2.2., we obtain immediately:

COROLLARY 1.2.3. Each plant $P \in M(\mathbb{R}(s))$ is stabilizable, and any stabilizing controller C has an r.b.f. and an l.b.f.

Now we have seen that each plant $P \in M(\mathbb{R}(s))$ is stabilizable, we want to know how this can be done. To this end, we will now parametrize all stabilizing compensators of a certain plant P.

Let S(P) denote the set of all compensators C that stabilize P. Assume that (D,N) and $(\tilde{N},\tilde{D})$ are an r.b.f. and an l.b.f. of P respectively. Let $C_0 \in S(P)$, and (Z,Y) and $(\tilde{Y},\tilde{Z})$ be an l.b.f. and an r.b.f. of C_0 respectively, such that:

$$\begin{bmatrix} -Z & Y \\ \tilde{D} & \tilde{N} \end{bmatrix} \begin{bmatrix} -N & \tilde{Y} \\ D & \tilde{Z} \end{bmatrix} = \begin{bmatrix} I & 0 \\ 0 & I \end{bmatrix} \qquad (1.10)$$

From (1.10) we obtain for any m×n transfermatrix $R \in M(\mathbb{R}H_\infty)$:

$$\begin{bmatrix} -Z-R\tilde{D} & Y-R\tilde{N} \\ \tilde{D} & \tilde{N} \end{bmatrix} \begin{bmatrix} -N & \tilde{Y}-NR \\ D & \tilde{Z}+DR \end{bmatrix} = \begin{bmatrix} I & 0 \\ 0 & I \end{bmatrix} \qquad (1.11)$$

With formula (1.11) it is not very difficult to prove the following parametrization theorem (for an exact proof, see [15, p.108]).

THEOREM 1.2.4. Suppose $P \in M(\mathbb{R}(s))$, and let (D,N) and $(\tilde{N},\tilde{D})$ be any r.b.f. and l.b.f. of P. Select matrices $Y,Z,\tilde{Y},\tilde{Z} \in M(\mathbb{R}H_\infty)$ such that YD + ZN = I, $\tilde{D}\tilde{Y} + \tilde{N}\tilde{Z} = I$. Then:

$$\begin{aligned} S(P) &= \{ \ (Y-R\tilde{N})^{-1}(Z+R\tilde{D}) \ \mid \ R \in M(\mathbb{R}H_\infty), \ |Y-R\tilde{N}| \neq 0 \ \} \\ &= \{ \ (\tilde{Z}+DR)(\tilde{Y}-NR)^{-1} \ \mid \ R \in M(\mathbb{R}H_\infty), \ |\tilde{Y}-NR| \neq 0 \ \} \end{aligned} \qquad (1.12)$$

Finally we prove the following result:

THEOREM 1.2.5. Suppose $P \in M(\mathbb{R}(s))$ and let (D,N) and $(\tilde{N},\tilde{D})$ be any r.b.f. and l.b.f. of P respectively. Let $C_0 \in S(P)$ and (Z,Y) and $(\tilde{Y},\tilde{Z})$ an l.b.f. and an r.b.f. of C_0 respectively, such that (1.10) holds. Then we have for any n×m transfermatrix $R \in M(\mathbb{R}H_\infty)$:

1) $|Y-R\tilde{N}| \neq 0$ if and only if $|\tilde{Y}-NR| \neq 0$;

2) If $|Y-R\tilde{N}| \neq 0$, then

$$(Y-R\tilde{N})^{-1}(Z+R\tilde{D}) = (\tilde{Z}+DR)(\tilde{Y}-NR)^{-1} \tag{1.13}$$

PROOF 1) Assume $|Y - R\tilde{N}| \neq 0$. According to theorem 1.2.5., the compensator $C = (Y-R\tilde{N})^{-1}(Z+R\tilde{D})$ is an element of $S(P)$. Let (Y_c, Z_c) be an r.b.f. of C. Again using theorem 1.2.4., we see that there exists an $R_c \in M(\mathbb{R}H_\infty)$ such that $|\tilde{Y} - NR_c| \neq 0$ and $(Y_c, Z_c) = (\tilde{Y}-NR_c, \tilde{Z}+DR_c)$. So

$$(Y-R\tilde{N})^{-1}(Z+R\tilde{D}) = Z_c Y_c^{-1} = (\tilde{Z}+DR_c)(\tilde{Y}-NR_c)^{-1}. \tag{1.14}$$

Pre–multiplying (1.14) with $(Y-R\tilde{N})$ and post–multiplying it by $(\tilde{Y}-NR_c)$ gives

$$(Z+R\tilde{D})(\tilde{Y}-NR_c) = (Y-R\tilde{N})(\tilde{Z}+DR_c),$$

so

$$Z\tilde{Y}-ZNR_c+R\tilde{D}\tilde{Y}-R\tilde{D}NR_c = Y\tilde{Z}+YDR_c-R\tilde{N}\tilde{Z}-R\tilde{N}DR_c.$$

Rearranging gives:

$$R(\tilde{D}\tilde{Y}+\tilde{N}\tilde{Z}) - (ZN+YD)R_c = (Y\tilde{Z}-Z\tilde{Y}) - R(\tilde{N}D-\tilde{D}N)R_c.$$

But by formula (1.10) we know that $(\tilde{D}\tilde{Y}+\tilde{N}\tilde{Z}) = I$, $(ZN+YD) = I$, $Y\tilde{Z}-Z\tilde{Y} = 0$ and $\tilde{N}D-\tilde{D}N = 0$, so $R-R_c = 0$, and $R = R_c$. Hence $|\tilde{Y} - NR| \neq 0$.

The converse can be proved completely analogously.

2) The proof of the second statement follows already from the arguments above. Combination of (1.14) and the fact that $R = R_c$ immediately gives (1.13). □

1.3. Robust stabilization

In this section we formulate the problem of robust stabilization, described in the introduction, more precisely. Suppose we have a sequence $\{P_\lambda\}$ of systems and a sequence $\{C_\lambda\}$ of controllers, parametrized by λ taking values in a metric space Λ. Suppose $H(P_0,C_0)$ is stable. The *central question* is: *when will* $H(P_\lambda,C_\lambda)$ *also be stable as* λ *is sufficiently close to 0, and* $H(P_\lambda,C_\lambda) \longrightarrow H(P_0,C_0)$ *as* $\lambda \longrightarrow 0$. Here the space Λ of parameters can arise from perturbations, disturbances, modelling errors, measurement errors, etc.

In practice this means the following. P_0 is the nominal system or a mathematical model (P_m in the introduction), which approximately describes the unknown real physical system. C_0 is the nominal (ideal) controller, designed according to the nominal plant P_0. The nominal system P_0 and compensator C_0 form a stable pair, i.e. $H(P_0,C_0)$ is stable. Now we hope that in practice the real physical system P_λ, which is close to P_0, and the real controller C_λ, which is close to C_0, also form a stable pair, and that $H(P_\lambda,C_\lambda)$ is close to $H(P_0,C_0)$. This is the problem of *robustness of feedback stabilization*, or simply *robust stabilization*.

We even want to go a step further. In section 1.2. we saw that there is not one stabilizing controller for a plant P, but a whole set: $S(P)$. Now we are interested in that controller C that meets the robustness–criterion the best. In other words, we want to find the optimally robust controller C. That is the controller from the set $S(P)$, that not only stabilizes P, but also the largest neighborhood of P.

To make the last statements more precise we have to introduce a topology or a metric for plants, in order to describe the distance from P_0 to P_λ, and from C_0 to C_λ. The operator–norm is not applicable for this purpose, because this norm is not defined for unstable systems. However, we require that this topology is compatible with the robustness of stabilization, in the sense that perturbation (P_λ,C_λ) from (P_0,C_0) is a stable pair, and $H(P_\lambda,C_\lambda)$ is close to $H(P_0,C_0)$ if P_λ is close to P_0 and C_λ is close to C_0 in the topology. The following topology meets this requirement.

Let $\mathbb{R}^{n*m}(s)$ denote the set of all $n \times m$ matrices with elements in $\mathbb{R}(s)$. First we define a basic neighborhood N of $P_0 \in \mathbb{R}^{n*m}(s)$:

$$N := N(P_0,C_0,\varepsilon) := \{\ P \in \mathbb{R}^{n*m}(s) \mid H(P,C_0) \text{ is stable}$$
$$\text{and } \| H(P,C_0) - H(P_0,C_0) \| < \varepsilon\ \} \qquad (1.15)$$

where C_0 is one of the stabilizing controllers of P_0. By varying ε over $\mathbb{R}_+$, C_0 over $S(P_0)$, and P_0 over $\mathbb{R}^{n*m}(s)$, we obtain a collection of basic neighborhoods, which forms a basis for a topology T over $\mathbb{R}^{n*m}(s)$.

Although this topology perfectly describes the robustness of feedback stabilization, unfortunately it has too little structure to be studied from this definition. That is why, in the next chapter, we will introduce another topology, the gap–topology, which in the end will appear to be identical to the topology T on $\mathbb{R}^{n*m}(s)$.

2. THE GAP–TOPOLOGY

In this section we introduce a topology for plants: the gap–topology. First we define the gap between two closed subspaces of a Banach–space X. With the aid of this concept, we can describe the gap–topology for plants. Then we first quote some results about the gap–topology from other authors, before we derive a necessary and sufficient condition for robust stabilization. Finally we show how the gap–topology can be metrized, and how this so called gap–metric can be computed. In the last section we shortly glance at the graph–topology, another topology for plants, suggested by Vidyasagar in [15, sec.7.2], and compare this to our gap–topology.

2.1. The gap between two closed subspaces

Let X be a Banach–space and φ and ψ two closed subspaces in X. We define

$$S_\varphi := \{\ x \in \varphi\ |\ \|\ x\ \| = 1\ \}$$

Now the *directed gap* from φ to ψ is given by

$$\vec{\delta}(\varphi,\psi) := \sup_{x \in S_\varphi}\ \inf_{y \in \psi}\ \|\ x{-}y\ \| \tag{2.1}$$

In order to measure the "distance" between two closed subspaces, the *gap between φ and ψ* is now defined as:

$$\delta(\varphi,\psi) := \max\ \{\ \vec{\delta}(\varphi,\psi)\ ,\ \vec{\delta}(\psi,\varphi)\ \} \tag{2.2}$$

Moreover, we define

$$\vec{\delta}(0,\varphi) := 0.$$

From these definitions and the fact that $\vec{\delta}(\varphi,0) = 1$ if and only if $\varphi \neq 0$, we get directly the following relations:

$$\delta(\varphi,\psi) = 0 \text{ iff } \varphi = \psi; \qquad\qquad \delta(\varphi,\psi) = \delta(\psi,\varphi);$$
$$0 \leq \delta(\varphi,\psi) \leq 1$$

Remark that the gap $\delta(.,.)$ is in general not a metric on the space of all closed subspaces of X, because the triangle inequality is not always satisfied.

2.2. The gap—topology

Before we introduce the gap—topology, we first look at the transfermatrix P of a plant P in a slightly different way. Suppose $P \in \mathbb{R}^{n*m}(s)$. Then P transforms an m—dimensional input to an n—dimensional output. In our case the input— and output—space are, apart from dimension, the same: the Banach—space H_2.

DEF.2.2.1. The *Hardy—space* H_2^n is the space of all functions x(s) which are analytic in Re s > 0, take values in $\mathbb{C}^n$, and satisfy the uniform square—integrability condition:

$$\| x \|_2 := [\sup_{\xi > 0} (2\pi)^{-1} \int_{-\infty}^{\infty} \| x(\xi + i\omega) \|^2 \, d\omega]^{1/2} < \infty \qquad (2.3)$$

(2.3) defines a norm on H_2^n, with this norm H_2^n is a Banach—space.

(Because H_2^n is a closed subspace of the Hilbert—space L_2^n, H_2^n is a Hilbert—space itself.)

Each transfermatrix $P(.) \in \mathbb{R}^{n*m}(s)$ now induces an operator P: the domain Dom(P) of P consists of all elements x(.) in H_2^m such that the product $P(.)x(.)$ is in H_2^n, the action of P on $x(.) \in$ Dom(P) is defined as $P(.)x(.)$. In this way we see a transfermatrix $P(.) \equiv \mathbb{R}^{n*m}(s)$ as an operator P, mapping a subspace of H_2^m into H_2^n.

For the operator P, we can define the *graph* of P as:

$$G(P) := \{ (x, Px) \mid x \in \text{Dom}(P) \}. \qquad (2.4)$$

In [18, p.850] Zhu proved that for each plant—compensator pair $(P,C) \in M(\mathbb{R}(s))$, such that H(P,C) is stable, the induced operators P and C are closed. Since we know (corollary 1.2.3.) that each $P \in M(\mathbb{R}(s))$ is stabilizable, it follows immediately that for each $P \in \mathbb{R}^{n*m}(s)$, the induced operator P is closed. So G(P) is a closed subspace of $(X^m \times X^n)$.

Now we can introduce the gap between two systems P_1 and P_2. Suppose $P_1, P_2 \in \mathbb{R}^{n*m}(s)$. The *gap between* P_1 *and* P_2 is defined as the gap between their graphs respectively, i.e.

$$\delta(P_1, P_2) := \delta(G(P_1), G(P_2)). \qquad (2.5)$$

From this formula follows immediately that $\delta(P_1, P_2) = 0$ if and only if $P_1 = P_2$.

A *basic neighborhood* of $P_0 \in \mathbb{R}^{n*m}(s)$ we define as:

$$N(P_0,\varepsilon) := \{ \ P \in \mathbb{R}^{n*m}(s) \ | \ \delta(P,P_0) < \varepsilon \ \}.$$

By varying ε over $(0,1]$ and varying P_0 over $\mathbb{R}^{n*m}(s)$, we obtain a collection of basic neighborhoods. This collection is a basis for a topology on $\mathbb{R}^{n*m}(s)$, which we call the *gap–topology*.

We now give some properties of the gap–topology, quoted from other authors. The lemma's 2.2.1.–2.2.4. come from Kato ([9, pp.197–206]), lemma 2.2.5. from Zhu ([18, p. 852]).

LEMMA 2.2.1. If P_0 is an n×m matrix, $P_0 \in M(\mathbb{R}H_\infty)$, and $P \in \mathbb{R}^{n*m}(s)$ satisfies

$$\delta(P,P_0) < (1 + \| P_0 \|^2)^{-1/2}$$

then $P \in M(\mathbb{R}H_\infty)$.

So the set of n×m matrices in $M(\mathbb{R}H_\infty)$ is an open subset of $\mathbb{R}^{n*m}(s)$ in the gap–topology. As a consequence, any system is stable when it is sufficiently close to a stable one.

THEOREM 2.2.2. On $M(\mathbb{R}H_\infty)$ is the gap–topology identical to the topology induced by the operator–norm.

LEMMA 2.2.3. Let $P_1, P_2 \in \mathbb{R}^{n*m}(s)$, and P_0 an n×m matrix in $M(\mathbb{R}H_\infty)$. Then:

$$\delta(P_1+P_0, P_2+P_0) \leq 2 \ (1 + \| P_0 \|^2) \ \delta(P_1, P_2). \qquad (2.6)$$

LEMMA 2.2.4. If $P_1, P_2 \in \mathbb{R}^{n*n}(s)$ are invertible, then

$$\delta(P_1^{-1}, P_2^{-1}) = \delta(P_1, P_2). \qquad (2.7)$$

LEMMA 2.2.5. Suppose that $P_i \in \mathbb{R}^{n*m}(s)$ (i=1,2) have the following diagonal form:

$$P_i = \begin{bmatrix} P_i^1 & 0 \\ 0 & P_i^2 \end{bmatrix} \quad (i=1,2). \qquad (2.8)$$

Then:

$$\max \{ \ \delta(P_1^1,P_2^1),\delta(P_1^2,P_2^2) \ \} \le \delta(P_1,P_2) \le \delta(P_1^1,P_2^1) + \delta(P_1^2,P_2^2).$$

From lemma 2.2.5. we immediately derive (the proof is obvious):

COROLLARY 2.2.6. Let $\{P_\lambda\}$ be a sequence of systems in $\mathbb{R}^{m*n}(s)$, with the diagonal form:

$$P_\lambda = \begin{bmatrix} P_\lambda^1 & 0 \\ 0 & P_\lambda^2 \end{bmatrix}.$$

Then, $\delta(P_\lambda,P_0) \longrightarrow 0$ (as $\lambda \longrightarrow 0$) if and only if $\delta(P_\lambda^1,P_0^1) \longrightarrow 0$ and $\delta(P_\lambda^2,P_0^2) \longrightarrow 0$ simultaneously (as $\lambda \longrightarrow 0$).

We will use this so called diagonal product property in the next section.

2.3. A necessary and sufficient condition for robust stabilization

In this section we apply the gap–topology to the problem of robust stabilization. We show that the gap–topology is compatible with this problem, and is identical to the topology T, given in section 1.3.

THEOREM 2.3.1. Let $\{P_\lambda\} \subset \mathbb{R}^{n*m}(s)$ be a sequence of systems, and $\{C_\lambda\} \subset \mathbb{R}^{m*n}(s)$ a sequence of controllers. Then:

$$\delta(H(P_\lambda,C_\lambda),H(P_0,C_0)) \longrightarrow 0 \quad (\lambda \longrightarrow 0)$$

iff

$$\delta(P_\lambda,P_0) \longrightarrow 0 \quad (\lambda \longrightarrow 0); \qquad \delta(C_\lambda,C_0) \longrightarrow 0 \quad (\lambda \longrightarrow 0)$$

simultaneously.

PROOF It is not difficult to show that $H_\lambda := H(P_\lambda,C_\lambda)$ can be written as:

$$H_\lambda = (I + FG_\lambda)^{-1}$$

with

$$F := \begin{bmatrix} 0 & I \\ -I & 0 \end{bmatrix} \qquad\qquad G_\lambda := \begin{bmatrix} C_\lambda & 0 \\ 0 & P_\lambda \end{bmatrix}$$

(see for example [15, pp.100–101])

According to lemma 2.2.4. we have

$$\delta(H_\lambda, H_0) = \delta((I+FG_\lambda)^{-1} , (I+FG_0)^{-1})$$
$$= \delta((I+FG_\lambda) , (I+FG_0)). \tag{2.9}$$

Now the application of (2.6) (two times) gives:

$$\tfrac{1}{4} \, \delta(FG_\lambda, FG_0) =$$
$$= \tfrac{1}{4} \, \delta((I+FG_\lambda)+(-I) , (I+FG_0)+(-I)) \leq$$
$$\leq \delta((I+FG_\lambda) , (I+FG_0)) \leq 4 \, \delta(FG_\lambda, FG_0). \tag{2.10}$$

(2.9) and (2.10) together yield

$$\tfrac{1}{4} \, \delta(FG_\lambda, FG_0) \leq \delta(H_\lambda, H_0) \leq 4 \, \delta(FG_\lambda, FG_0). \tag{2.11}$$

Analogous to 2.2.5. it can be proved that lemma 2.2.5. still holds when P_i (i=1,2) is defined as

$$P_i := \begin{bmatrix} 0 & P_i^1 \\ P_i^2 & 0 \end{bmatrix} \quad (i=1,2).$$

This implies that

$$\max \{\delta(P_\lambda, P_0), \delta(C_\lambda, C_0)\} \leq \delta(FG_\lambda, FG_0) \leq \delta(P_\lambda, P_0) + \delta(C_\lambda, C_0).$$

Combination of this last formula with (2.11) yields

$$\tfrac{1}{4} \, \max \{\delta(P_\lambda, P_0), \delta(C_\lambda, C_0)\} \leq \delta(H_\lambda, H_0) \leq 4 \, [\delta(P_\lambda, P_0) + \delta(C_\lambda, C_0)]$$
$$\tag{2.12}$$

From (2.12) our claim follows immediately. □

With help of theorem 2.3.1. we can now give a necessary and sufficient condition for robustness of feedback stabilization.

COROLLARY 2.3.2. Consider the feedback system in fig.(1.3). Suppose P_0 and C_0 are the nominal system and controller respectively, and $H(P_0, C_0)$ is stable. Let $\{P_\lambda\}_{\lambda > 0}$ and $\{C_\lambda\}_{\lambda > 0}$ be families of systems and controllers respectively. Then:

$$\exists \; \lambda_0 > 0: \; \forall \; \lambda < \lambda_0: \; H(P_\lambda, C_\lambda) \text{ is stable,}$$

and

$$\| \; H(P_\lambda, C_\lambda) - H(P_0, C_0) \; \| \longrightarrow 0 \quad (\lambda \longrightarrow 0) \tag{2.13}$$

if and only if

$$\delta(P_\lambda, P_0) \longrightarrow 0 \quad (\lambda \longrightarrow 0) \qquad\qquad \delta(C_\lambda, C_0) \longrightarrow 0 \quad (\lambda \longrightarrow 0)$$

simultaneously.

PROOF "$\Rightarrow$" Suppose $H(P_\lambda, C_\lambda)$ is stable for $\lambda < \lambda_0$, and $\| \; H(P_\lambda, C_\lambda) - H(P_0, C_0) \; \| \longrightarrow 0$ ($\lambda \longrightarrow 0$). $H(P_\lambda, C_\lambda)$ is stable. According to theorem 2.2.2., the gap–topology is identical to the topology induced by the operator norm. Therefore we have: $\delta(H(P_\lambda, C_\lambda), H(P_0.C_0)) \longrightarrow 0$ ($\lambda \longrightarrow 0$). Now application of theorem 2.3.1. gives the necessity.

"$\Leftarrow$" Suppose $\delta(P_\lambda, P_0) \longrightarrow 0$ ($\lambda \longrightarrow 0$) and $\delta(C_\lambda, C_0) \longrightarrow 0$ ($\lambda \longrightarrow 0$) simultaneously. Then by theorem 2.3.1. we have $\delta(H(P_\lambda, C_\lambda), H(P_0, C_0)) \longrightarrow 0$ ($\lambda \longrightarrow 0$). According to lemma 2.2.1. $H(P_\lambda, C_\lambda)$ is stable as λ is sufficiently close to 0 (because then $H(P_\lambda, C_\lambda)$ is close to $H(P_0, C_0)$, which is stable). Again application of theorem 2.2.2. gives that $\delta(H(P_\lambda, C_\lambda), H(P_0, C_0)) \longrightarrow 0$ ($\lambda \longrightarrow 0$) now implies $\| \; H(P_\lambda, C_\lambda) - H(P_0, C_0) \; \| \longrightarrow 0$ ($\lambda \longrightarrow 0$). □

Finally we show that the gap–topology on $\mathbb{R}^{n*m}(s)$ is identical to the topology T, defined in section 1.3. In this toplogy, a basic neighborhood of $P \in \mathbb{R}^{n*m}(s)$ was defined as

$$N(P_0, C_0, \varepsilon) := \{ \; P \in \mathbb{R}^{n*m}(s) \; | \; H(P, C_0) \text{ is stable and}$$
$$\| \; H(P, C_0) - H(P_0, C_0) \; \| < \varepsilon \; \}$$

with C_0 one of the stabilizing compensators of P_0.

Now suppose the sequence $\{P_\lambda\} \subset \mathbb{R}^{n*m}(s)$ converges to $P_0 \in \mathbb{R}^{n*m}(s)$ in the topology T. Then we have that $H(P_\lambda, C_0)$ is stable when λ is sufficiently close to 0, and $\| \; H(P_\lambda, C_0) - H(P_0, C_0) \; \| \longrightarrow 0$ ($\lambda \longrightarrow 0$). So according to corollary 2.3.2., $\{P_\lambda\}$ converges to P_0 in the gap–topology also.

Conversely, suppose $\{P_\lambda\} \subset \mathbb{R}^{n*m}(s)$ converges to $P_0 \in \mathbb{R}^{n*m}(s)$ in the gap–topology. Suppose $H(P_0, C_0)$ is stable. Then, by corollary 2.3.2., we know that $H(P_\lambda, C_0)$ is stable when λ is sufficiently close to 0, and $\| \; H(P_\lambda, C_0) - H(P_0, C_0) \; \| \longrightarrow 0$ ($\lambda \longrightarrow 0$). This means that $\{P_\lambda\}$ converges to P_0 in the topology T.

2.4. The gap–metric

In section 2.1. we saw that the gap δ between two closed subspaces is in general not a metric. In this section we show that in our case, where the space H_2 of inputs and outputs is a Hilbert–space, the gap δ between two plants is in fact a metric. We also give a method to compute this gap–metric in practice.

Suppose $P_1, P_2 \in \mathbb{R}^{n*m}(s)$, and we regard P_1 and P_2 as operators from a subspace of the Hilbert–space H_2^m to the Hilbert–space H_2^n. Let $\Pi(P_i)$ (i=1,2) denote the orthogonal projection from $H_2^m \times H_2^n$ onto the graph $G(P_i)$ of P_i (i=1,2). Then it is not difficult to see that

$$
\begin{aligned}
\delta(P_1, P_2) \quad &= \sup_{x \in S_{G(P_1)}} \quad \inf_{y \in G(P_2)} \quad \| x - y \| = \\
&= \sup_{x \in S_{G(P_1)}} \quad \| (I - \Pi(P_2))x \| = \\
&= \sup_{x \in S_{G(P_1)}} \quad \| (I - \Pi(P_2))\Pi(P_1)x \| = \\
&= \sup_{x \in H_2^m * H_2^n, \ \|x\|=1} \quad \| (I - \Pi(P_2))\Pi(P_1)x \| = \\
&= \| (I - \Pi(P_2))\Pi(P_1) \|.
\end{aligned}
\tag{2.14}
$$

With this formula it is shown (see [9, pp. 56–58], or [10, p.205] for a direct proof) that

$$
\delta(P_1, P_2) = \| \Pi(P_1) - \Pi(P_2) \|.
\tag{2.15}
$$

From formula (2.15) it is clear that $\delta(.,.)$ defines a metric for plants. We call this metric the *gap–metric*.

To calculate the gap–metric, we need to find a representation of $\Pi(P)$ for $P \in M(\mathbb{R}(s))$. Such a representation can be given with help of Bezout factorizations. However, we need the following preliminary lemma (for a proof, see [20, p.55]).

LEMMA 2.4.1. Let $P \in \mathbb{R}^{n*m}(s)$, and suppose (D,N) is an r.b.f. of P. Let D^* and N^* denote the adjoint operators of D and N (in L_2–sense) respectively (so $D^* = D^T(-s)$, $N^* = N^T(-s)$). Then

$$
S := D^*D + N^*N
$$

has a bounded inverse.

Analogously, if $(\tilde{N}, \tilde{D})$ is an l.b.f. of P, then $S := (\tilde{D}\tilde{D}^* + \tilde{N}\tilde{N}^*)$ has a bounded inverse. With this knowledge the following theorem is not difficult to prove. In fact, it is a direct consequence of lemma 3.14 in [3, p.704].

THEOREM 2.4.2. Let $P \in \mathbb{R}^{n*m}(s)$, and (D,N) and $(\tilde{N}, \tilde{D})$ an r.b.f. and an l.b.f. of P respectively. Then:

$$\Pi(P) = \left[\begin{array}{c} D \\ N \end{array} \right] (D^*D + N^*N)^{-1} (D^*, N^*)$$

$$= I - \left[\begin{array}{c} -\tilde{D}^* \\ \tilde{N}^* \end{array} \right] (\tilde{D}\tilde{D}^* + \tilde{N}\tilde{N}^*)^{-1} (-\tilde{D}, \tilde{N}). \qquad (2.16)$$

With help of the formula's (2.15) and (2.16) it is in principle possible to calculate the gap between two plants.

Another, probably better, method to compute the gap–metric was introduced by Georgiou (see [5, p.254]). In this article he proved the following theorem.

THEOREM 2.4.3. Let $P_1, P_2 \in \mathbb{R}^{n*m}(s)$, and (D_1, N_1) and (D_2, N_2) *normalized* right–Bezout factorizations of P_1 and P_2 respectively. Then:

$$\vec{\delta}(P_1, P_2) = \inf_{Q \in M(H_\infty)} \left\| \left[\begin{array}{c} D_1 \\ N_1 \end{array} \right] - \left[\begin{array}{c} D_2 \\ N_2 \end{array} \right] Q \right\|. \qquad (2.17)$$

An exact definition of normalized right–Bezout factorizations will be given in chapter 4. We here only mention that it is an r.b.f. with the extra property

$$D^*D + N^*N = I$$

and that such a factorization for a $P \in \mathbb{R}^{n*m}(s)$ always exists (see [15, p.262]).

From theorem 2.4.3. we see that the computation of the gap between two plants can be reduced to the computation of two infima as given in formula (2.17). For $P \in M(\mathbb{R}_p(s))$ such infima can be computed with the method of Francis, described in [4, ch.8]. In a later article [6, pp.5–6] Georgiou applied this method to this problem, and worked it out. So this later article gives a (theoretical) algorithm to compute the gap–metric. For computer purposes this algorithm is implemented in the MATLAB–function gap. For the details of this implementation we refer to appendix A.

2.5. An other topology for plants

In this last section of chapter 2, we want, for the sake of completeness, mention an other topology for plants: the graph–topology. This topology was introduced by Vidyasagar in [15, sec.7.2.]. For a definition and properties, we therefore refer to this book. Zhu however proved (see [18, pp. 853–854]) that on $\mathbb{R}^{n*m}(s)$ the gap– and graph–topology are identical.

Although the gap– and the graph–topology are, in our case, the same, the gap–topology has one great advantage: the gap–metric is much easier to compute than the graph–metric. In the last section we mentioned the method of Georgiou, that really solves the computation problem for the gap–metric. The graph–metric, on the other hand, is much more difficult to compute.

Another advantage of the gap–topology, which is not so important to us, is that when one works with more general systems, the gap–topology is defined for a greater class of systems than the graph–topology. Zhu even proved that in general the class of systems, for which the graph–topology is defined, is a subset of the class of systems, for which the gap–topology is defined (see [18, p.853]).

Because of these two reasons, we decided to work only with the gap–topology.

3. SUFFICIENT CONDITIONS FOR ROBUST BIBO STABILIZATION

In this chapter we derive sufficient conditions for robustness of feedback stabilization. In the first section we study the relation between the gap–metric and Bezout factorizations. With the results of this section we are able to derive several bounds which guarantee the stability of a perturbed feedback system, if the perturbations of the system and the compensator are within these bounds. This is the content of the second section.

3.1. The gap–metric and Bezout factorizations

In this section we dig out, as said before, the relation between the gap–metric and Bezout factorizations. Many of the results presented here, will turn out to be useful in the next section.

We start with the following lemma.

LEMMA 3.1.1. Assume $P \in \mathbb{R}^{n*m}(s)$, and regard P as an operator mapping a subspace of H_2^m to H_2^n. Denote the graph of P by $G(P)$. Let $D \in B(H_2^m)$ and $N \in B(H_2^m, H_2^n)$. Then:

$\qquad$ (D,N) is an r.b.f. of P

$\Leftrightarrow$

$\qquad$ (1) $G(P) = \text{Range} \begin{bmatrix} D \\ N \end{bmatrix} = \{ (Dz, Nz) \mid z \in H_2^m \}$ $\qquad\qquad\qquad$ (3.1)

$\qquad$ (2) Ker $D = \{0\}$

PROOF "$\Rightarrow$" Suppose (D,N) is an r.b.f. of P. Because D is invertible, clearly we have Ker D = {0}. So we only have to prove (1).

$\qquad$ Suppose $y \in G(P)$. Then $\exists\; x \in \text{Dom}(P) \subset H_2^m : y = (x, Px) = (x, ND^{-1}x)$. Because D is invertible, there exists a $z \in H_2^m$ such that $x = Dz$. This we can see as follows. Since (D,N) is an r.b.f. of P, there exist bounded operators Y and Z (Y and Z are matrices in $M(\mathbb{R}H_\infty)$) such that $YN + ZD = I$. Define $z := (XP+Y)x$. Then clearly $z \in H_2^m$ and we have:

$$Dz = D(XP+Y)x = D(XND^{-1} + YDD^{-1})x = D(XN+YD)D^{-1}x = DD^{-1}x = x$$

So, indeed we have $x = Dz$ and it follows immediately that:

$$y = (x, ND^{-1}x) = (Dz, ND^{-1}Dz) = (Dz, Nz) \in \text{Range} \begin{bmatrix} D \\ N \end{bmatrix}.$$

$\qquad$ Now let $y \in \text{Range} \begin{bmatrix} D \\ N \end{bmatrix}$. Then $\exists\; z \in H_2^m : y = (Dz, Nz)$. Define $x := Dz$. Then $x \in \text{Dom}(P)$ because $Px = PDz = ND^{-1}Dz = Nz$ is bounded (N is bounded). So $(x, Px) \in G(P)$ and we have $y = (Dz, Nz) = (x, Px) \in G(P)$.

"⇐" Now suppose (1) and (2) hold. Assume (D_1,N_1) is an r.b.f. of P. Because $P \in \mathbb{R}^{n*m}(s)$ such a pair exists. By the first part of the proof we know that

$$G(P) = \{ \ (D_1z,N_1z) \ | \ z \in H_2^m \ \}. \tag{3.2}$$

By (3.2) and (3.1) we know that for every $x \in H_2^m$ there exists an unique $y \in H_2^m$ such that

$$\left[\begin{array}{c} D \\ N \end{array} \right] x = \left[\begin{array}{c} D_1 \\ N_1 \end{array} \right] y$$

and vice versa.

Because (D_1,N_1) is an r.b.f. of P there exist bounded operators Y,Z (Y and Z are matrices in $M(\mathbb{R}H_\infty)$) such that $YD_1 + ZN_1 = I$. Therefore:

$$(YD + ZN)x = (Y,Z)\left[\begin{array}{c} D \\ N \end{array} \right] x = (Y,Z)\left[\begin{array}{c} D_1 \\ N_1 \end{array} \right] y = y.$$

So $U := (YD + ZN)$ maps H_2^m to H_2^m bijectively. Since $\left[\begin{array}{c} D \\ N \end{array} \right] = \left[\begin{array}{c} D_1 \\ N_1 \end{array} \right] U$, (D,N) must be an r.b.f. of P. □

The next lemma is an alternative version of a result in [10, p.206].

LEMMA 3.1.2. Let $P_1, P_2 \in \mathbb{R}^{n*m}(s)$. Then $\Pi(P_1)$ maps $G(P_2)$ bijectively onto $G(P_1)$ if and only if $\delta(P_1,P_2) < 1$.

In [9] Kato derived almost the same result (see [9, pp. 56–58, theorem 6.34]). His formulation of the theorem, however, gives rise to the following surprising result.

LEMMA 3.1.3. Let $P_1, P_2 \in \mathbb{R}^{n*m}(s)$. If $\delta(P_1,P_2) < 1$, then

$$\breve{\delta}(P_1,P_2) = \breve{\delta}(P_2,P_1) = \delta(P_1,P_2).$$

With help of lemma 3.1.1. and 3.1.2. we can now prove the following theorem.

THEOREM 3.1.4. Let $P_1, P_2 \in \mathbb{R}^{n*m}(s)$, and (D_1,N_1) an r.b.f. of P_1. Define

$$\left[\begin{array}{c} D_2 \\ N_2 \end{array} \right] := \Pi(P_2)\left[\begin{array}{c} D_1 \\ N_1 \end{array} \right]. \tag{3.3}$$

Then:

(D_2,N_2) is an r.b.f. of P_2

$\Leftrightarrow$

$\delta(P_1,P_2) < 1$.

PROOF "$\Leftarrow$" Suppose $\delta(P_1,P_2) < 1$. By lemma 3.1.2. we know that $\Pi(P_2)$ maps $G(P_1)$ onto $G(P_2)$ bijectively. So

$$
\begin{aligned}
G(P_2) &= \Pi(P_2)G(P_1) = \Pi(P_2) \text{ Range } \begin{bmatrix} D_1 \\ N_1 \end{bmatrix} = \\
&= \text{Range } \Pi(P_2) \begin{bmatrix} D_1 \\ N_1 \end{bmatrix} = \text{Range } \begin{bmatrix} D_2 \\ N_2 \end{bmatrix} = \\
&= \{ (D_2 z, N_2 z) \mid z \in H_2^m \}.
\end{aligned}
\tag{3.4}
$$

So $\begin{bmatrix} D_2 \\ N_2 \end{bmatrix}$ maps H_2^m surjectively onto $G(P_2)$.

But we also have ker $\begin{bmatrix} D_2 \\ N_2 \end{bmatrix} = \{0\}$. This we can see as follows. Suppose $\begin{bmatrix} D_2 \\ N_2 \end{bmatrix} z = 0$. Then $\Pi(P_2)\begin{bmatrix} D_1 \\ N_1 \end{bmatrix} z = 0$. And since $\Pi(P_2)$ maps $G(P_1)$ bijectively onto $G(P_2)$ we have $\begin{bmatrix} D_1 \\ N_1 \end{bmatrix} z = 0$. So $D_1 z = 0$, and because (D_1,N_1) is an r.b.f. we have, according to lemma 3.1.1., $z = 0$.

We draw the conclusion that $\begin{bmatrix} D_2 \\ N_2 \end{bmatrix}$ maps H_2^m bijectively onto $G(P_2)$. Now suppose $\begin{bmatrix} D \\ N \end{bmatrix}$ is an r.b.f. of P_2. Then $\begin{bmatrix} D \\ N \end{bmatrix}$ maps H_2^m bijectively onto $G(P_2)$. So there exists a bijective U, such that $\begin{bmatrix} D_2 \\ N_2 \end{bmatrix} = \begin{bmatrix} D \\ N \end{bmatrix} U$, and (D_2,N_2) is an r.b.f. of P_2.

"$\Rightarrow$" Suppose (D_2,N_2) is an r.b.f. of P_2. Then we know by lemma 3.1.1.

$$
\begin{aligned}
G(P_2) &= \text{Range } \begin{bmatrix} D_2 \\ N_2 \end{bmatrix} = \text{Range } (\Pi(P_2)\begin{bmatrix} D_1 \\ N_1 \end{bmatrix}) = \\
&= \Pi(P_2) \text{ Range } \begin{bmatrix} D_1 \\ N_1 \end{bmatrix} = \Pi(P_2) \, G(P_1).
\end{aligned}
$$

So $\Pi(P_2)$ maps $G(P_1)$ onto $G(P_2)$ surjectively.

Now, let $x \in G(P_1)$ and suppose $\Pi(P_2)x = 0$. Since $x \in G(P_1)$ and (D_1,N_1) is an r.b.f. of P_1, $x \in$ Range $\begin{bmatrix} D_1 \\ N_1 \end{bmatrix}$, so there exists a $z \in H_2^m$ such that $x = \begin{bmatrix} D_1 \\ N_1 \end{bmatrix} z$. So $\Pi(P_2)\begin{bmatrix} D_1 \\ N_1 \end{bmatrix} z = 0$, and applying (3.3) we get $\begin{bmatrix} D_2 \\ N_2 \end{bmatrix} z = 0$. So $D_2 z = 0$ and because (N_2,D_2) is an r.b.f. of P_2, this immediately yields $z = 0$. So $x = 0$, and $\Pi(P_2)$ maps $G(P_1)$ injectively onto $G(P_2)$.

Finally we conclude that $\Pi(P_2)$ maps $G(P_1)$ bijectively onto $G(P_2)$ and, according to lemma 3.1.2. we have $\delta(P_1,P_2) < 1$. $\square$

It is also possible to derive an analogous result as theorem 3.1.4. for left–Bezout factorizations. We have the following theorem (for a proof we refer to [20, p.57]).

THEOREM 3.1.5. Suppose $P_1,P_2 \in \mathbb{R}^{n*m}(s)$, and let $(\tilde{N}_1,\tilde{D}_1)$ an l.b.f. of P_1. Let $\tilde{N}_1^*$ and $\tilde{D}_1^*$ denote the Hilbert–adjoint operators of $\tilde{N}_1$ and $\tilde{D}_1$ respectively. Define

$$
\begin{bmatrix} -\tilde{D}_2 \\ \tilde{N}_2 \end{bmatrix} = (\Pi(P_2))^{\perp} \begin{bmatrix} -\tilde{D}_1^* \\ \tilde{N}_1^* \end{bmatrix}.
\tag{3.5}
$$

Then:

$$(\tilde{N}_2^*,-\tilde{D}_2^*) \text{ is an l.b.f. of } P_2$$

$\Leftrightarrow$

$$\delta(P_1,P_2) < 1.$$

Finally we mention the following result, which will be needed in the next chapter. With the tools developed in this section, it is not very difficult to derive this lemma.

LEMMA 3.1.6. Let (D_1,N_1) and (D_2,N_2) be r.b.f.'s of P_1 and $P_2 \in \mathbb{R}^{n*m}(s)$. Then:

$$\delta(P_1,P_2) < 1$$

$\Leftrightarrow$

$$N_1^*N_2 + D_1^*D_2 \text{ is bijective.}$$

PROOF According to theorem 3.1.4. $\delta(P_1,P_2) < 1$ if and only if $\Pi(P_1) \begin{bmatrix} D_2 \\ N_2 \end{bmatrix}$ is an r.b.f. of P_1. From formula (2.16) and the fact that $\begin{bmatrix} D_1 \\ N_1 \end{bmatrix}$ is an r.b.f. of P_1, we know that

$$
\Pi(P_1) = \begin{bmatrix} D_1 \\ N_1 \end{bmatrix}(D_1^*D_1+N_1^*N_1)^{-1}(D_1^*,N_1^*).
$$

So

$$
\Pi(P_1)\begin{bmatrix} D_2 \\ N_2 \end{bmatrix} = \begin{bmatrix} D_1 \\ N_1 \end{bmatrix}(D_1^*D_1+N_1^*N_1)^{-1}(D_1^*D_2+N_1^*N_2).
$$

Because $\begin{bmatrix} D_1 \\ N_1 \end{bmatrix}$ is already an r.b.f. of P_1, $\Pi(P_1)\begin{bmatrix} D_2 \\ N_2 \end{bmatrix}$ is an r.b.f. of P_1 if and only if $(D_1^*D_1+N_1^*N_1)^{-1}(D_1^*D_2+N_1^*N_2)$ is bijective. Since $(D_1^*D_1+N_1^*N_1)^{-1}$ is clearly bijective, we get as final result: $\delta(P_1,P_2) < 1$ if and only if $(D_1^*D_2+N_1^*N_2)$ is bijective. $\square$

3.2. Guaranteed bounds for robust stabilization

In this section we present two bounds which guarantee the stability of a perturbed feedback system, if the perturbations of the system and controller are within these bounds.

Suppose $P_0 \in \mathbb{R}^{n*m}(s)$ is the nominal system and $C_0 \in \mathbb{R}^{m*n}(s)$ the nominal controller, and assume that P_0 and C_0 form a stable pair, i.e. $H(P_0,C_0)$ is stable. Let $P_\lambda \in \mathbb{R}^{n*m}(s)$ and $C_\lambda \in \mathbb{R}^{m*n}(s)$ denote perturbed versions of P_0 and C_0 respectively.

The first bound we derive with help of the tools developed in chapter 2.

THEOREM 3.2.1. If $H(P_0,C_0)$ is stable and

$$\delta(P_\lambda,P_0) + \delta(C_\lambda,C_0) < \tfrac{1}{4} \, (1 + \parallel H(P_0,C_0) \parallel^2)^{-1/2} \qquad (3.6)$$

then $H(P_\lambda,C_\lambda)$ is stable.

PROOF From formula (2.12) it follows that

$$\delta(H(P_\lambda,C_\lambda),H(P_0,C_0)) \leq 4 \, [\delta(P_\lambda,P_0) + \delta(C_\lambda,C_0)].$$

So in this case we have

$$\delta(H(P_\lambda,C_\lambda),H(P_0,C_0)) < (1 + \parallel H(P_0,C_0) \parallel^2)^{-1/2}.$$

Now $H(P_0,C_0)$ forms a stable closed loop system, so according to lemma 2.2.1. we know that $H(P_\lambda,C_\lambda)$ is stable. □

To derive an other bound for robust stabilization, we need a few preliminary lemmas.

LEMMA 3.2.2. Suppose A is an operator, mapping a Hilbert–space X onto X. Assume $\parallel A \parallel < 1$. Then $(I+A)$ is invertible, and the inverse is bounded.

PROOF Define

$$T := \sum_{k=0}^{\infty} (-1)^k \, A^k.$$

Then

$$\parallel T \parallel = \parallel \sum_{k=0}^{\infty} (-1)^k \, A^k \parallel \leq \sum_{k=0}^{\infty} \parallel A \parallel^k \leq \frac{1}{1 - \parallel A \parallel}.$$

So T is bounded and it is not difficult to prove that $T = (I+A)^{-1}$:

$$(I+A)T = \sum_{k=0}^{\infty} (-1)^k A^k + \sum_{k=0}^{\infty} (-1)^k A^{k+1} =$$

$$= I + \sum_{k=1}^{\infty} (-1)^k A^k + \sum_{k=1}^{\infty} (-1)^{k-1} A^k = I.$$

$$T(I+A) = \sum_{k=0}^{\infty} (-1)^k A^k + \sum_{k=0}^{\infty} (-1)^k A^{k+1} = I. \quad \square$$

LEMMA 3.2.3. Suppose U_0 is an operator, mapping a Hilbert–space X onto X. Assume U_0 has a bounded inverse. Let U be an operator such that

$$\| U - U_0 \| < \| U_0^{-1} \|^{-1}.$$

Then U is also invertible, and the inverse is bounded.

PROOF $(U_0 - U) = U_0(I - U_0^{-1}U)$. If $\| U - U_0 \| < \| U_0^{-1} \|^{-1}$, we have

$$\| I - U_0^{-1}U \| = \| U_0^{-1}U_0 - U_0^{-1}U \| \leq \| U_0^{-1} \| \, \| U_0 - U \| < 1.$$

So $\| U_0^{-1}U - I \| < 1$. By lemma 3.2.2. we know that $I + U_0^{-1}U - I = U_0^{-1}U$ is invertible, and that the inverse is bounded. Since U_0 is a bounded invertible operator, it follows immediately that U is also invertible, with bounded inverse. $\square$

Now we return to our nominal system and controller P_0 and C_0. Let (D_0, N_0) an r.b.f. of P_0 and $(\tilde{N}_0, \tilde{D}_0)$ an l.b.f. of C_0. Denote

$$A_0 = \left[\begin{array}{c} D_0 \\ N_0 \end{array} \right], \qquad\qquad B_0 = (\tilde{D}_0, \tilde{N}_0). \qquad\qquad (3.7)$$

Define $U_0 := B_0 A_0$. We now have the following theorem (for a proof we refer to [15, p.105])

THEOREM 3.2.4. $H(P_0, C_0)$ is stable if and only if U_0 is a bounded operator which maps H_2^m bijectively onto H_2^m.

With help of this last theorem and the foregoing lemmas , we can prove theorem 3.2.5.

THEOREM 3.2.5. Let (D_0, N_0) an r.b.f. of P_0 and $(\tilde{N}_0, \tilde{D}_0)$ an l.b.f. of C_0, and assume $H(P_0, C_0)$ is stable. Denote

$$w = [\ \|\ A_0\ \|\ \|\ B_0\ \|\ \|\ U_0^{-1}\ \|\ ].$$

Then we have:
if

$$\delta(P_\lambda, P_0) + \delta(C_\lambda, C_0) < w^{-1} \tag{3.8}$$

then

$$H(P_\lambda, C_\lambda) \text{ is stable.}$$

PROOF First we show that the right–hand side of (3.8) is smaller or equal to one. We have $I = B_0 A_0 U_0^{-1}$ and so:

$$1 = \|\ I\ \| = \|\ B_0 A_0 U_0^{-1}\ \| \leq \|\ B_0\ \|\ \|\ A_0\ \|\ \|\ U_0^{-1}\ \| = w.$$

Thus $w \geq 1$, and $w^{-1} \leq 1$.

According to (3.8) we know that $\delta(P_\lambda, P_0) < 1$ and $\delta(C_\lambda, C_0) < 1$. We can now apply the theorems 3.1.4. and 3.1.5. to get an r.b.f. of P_λ and an l.b.f. of C_λ. Theorem 3.1.4. gives that (D_λ, N_λ) defined by

$$\begin{bmatrix} D_\lambda \\ N_\lambda \end{bmatrix} = \Pi(P_\lambda) \begin{bmatrix} D_0 \\ N_0 \end{bmatrix}$$

is an r.b.f. of P_λ, and theorem 3.1.5. that $(\tilde{N}_\lambda, \tilde{D}_\lambda)$ defined by

$$\begin{bmatrix} -\tilde{D}_\lambda^* \\ \tilde{N}_\lambda^* \end{bmatrix} = \Pi(C_\lambda)^\perp \begin{bmatrix} -\tilde{D}_0^* \\ \tilde{N}_0^* \end{bmatrix}$$

is an l.b.f. of C_λ. Analogous to (3.7) denote

$$A_\lambda = \begin{bmatrix} D_\lambda \\ N_\lambda \end{bmatrix}, \qquad\qquad B_\lambda = (\tilde{D}_\lambda, \tilde{N}_\lambda). \tag{3.9}$$

Then we have

$$\begin{aligned}
\|\ B_\lambda A_\lambda - B_0 A_0\ \| &= \\
&= \|\ B_\lambda A_\lambda - B_0 A_\lambda + B_0 A_\lambda - B_0 A_0\ \| = \\
&= \|\ (B_\lambda - B_0) A_\lambda + B_0 (A_\lambda - A_0)\ \| \leq
\end{aligned}$$

$$\leq \| A_\lambda \| \| B_\lambda - B_0 \| + \| A_\lambda - A_0 \| \| B_0 \| \leq$$
$$\leq \| A_\lambda \| \| (\Pi(C_\lambda) - \Pi(C_0))B_0 \| + \| (\Pi(P_\lambda) - \Pi(P_0))A_0 \| \| B_0 \|.$$

And since $A_\lambda = \Pi(P_\lambda)A_0$ we get

$$\| B_\lambda A_\lambda - B_0 A_0 \| \leq$$
$$\leq (\| \Pi(C_\lambda) - \Pi(C_0) \| + \| \Pi(P_\lambda) - \Pi(P_0) \|) \| A_0 \| \| B_0 \| =$$
$$= \| A_0 \| \| B_0 \| (\delta(C_\lambda, C_0) + \delta(P_\lambda, P_0)) <$$
$$< \| A_0 \| \| B_0 \| w^{-1} =$$
$$= \| U_0^{-1} \|^{-1}.$$

Now according to lemma 3.2.3. $U_\lambda = B_\lambda A_\lambda$ is invertible and the inverse is bounded. By theorem 3.2.4. it is then clear that $H(P_\lambda, C_\lambda)$ is stable. □

This last theorem will play a crucial role in the rest of this monograph.

4. OPTIMALLY ROBUST CONTROL

In this chapter we tackle the problem of optimally robust control. Given a nominal plant P_0, we want to design a compensator C that not only stabilizes P_0, but also plants P in the neighborhood of P_0. This neighborhood, stabilized by C, we want to make as large as possible. We call a compensator C_0 that stabilizes this largest neighborhood of P_0 an *optimally robust controller*.

To find an optimally robust controller, we suggest the following approach. In theorem 3.2.5. we found a guaranteed bound w^{-1} ($w = \| A_0 \| \| B_0 \| \|(B_0 A_0)^{-1}\|$) for robust stabilization. Now suppose P_0 is the nominal plant, and assume that C is a controller that stabilizes P_0. We also assume that the controller can be implemented exactly, so $C_\lambda = C$. Then theorem 3.2.5. becomes: if $\delta(P_\lambda, P_0) < w^{-1}$, then $H(P_\lambda, C)$ is stable. So all the plants in the ball around P_0 with radius w^{-1} are stabilized by C. Now w depends only upon P_0 and C, and P_0 is fixed. So by changing C in $S(P_0)$ we can change the value of w^{-1}. Now we are interested in a compensator $C_0 \in S(P_0)$ for which w^{-1} is as large as possible, because such a controller stabilizes all the plants in the largest ball around P_0 (the ball around P_0 with radius w^{-1}, where w^{-1} is as large as possible). In this way it is possible to find an, in a certain sense optimal, robust controller C_0.

In the second section of this chapter, we show that the approach suggested above, really leads to the solution of the problem of optimally robust control. Then, in the third section, we will derive a method to actually compute an optimally robust controller, and the radius of the ball around P_0, stabilized by this controller. But before we can do this, we have to introduce an other new concept: the idea of normalized Bezout factorizations. This will be the content of the first section of this chapter.

4.1. Normalized Bezout factorizations

Normalized Bezout factorizations, we mentioned them already in section 2.4., are a special kind of Bezout factorizations. In this section we give a formal definition and show by construction that each plant $P \in M(\mathbb{R}_p(s))$ has a normalized right- and left-Bezout factorization. The state-space realizations of these factorizations will be needed in chapter 5, where we develop an algorithm for the solution of the problem of optimally robust control for the case $P \in M(\mathbb{R}_p(s))$. Vidyasagar, however, showed that also in the more general case, a plant $P \in M(\mathbb{R}(s))$ has a normalized right- and left-Bezout factorization (see [15, p.262]).

Suppose $P \in \mathbb{R}^{n*m}(s)$, and let (D,N) be an r.b.f. of P. Define $D^*(s) := D^T(-s)$ and $N^*(s) := N^T(-s)$. So D^* and N^* are the adjoint operators of D and N respectively, seen as operators from L_2 to L_2. Then we have the following definition.

DEF.4.1.1. Let $P \in \mathbb{R}^{n*m}(s)$. We say $(D,N) \in M(\mathbb{R}H_\infty)$ is a *normalized right–Bezout factorization* (n.r.b.f.) of P if:

   1) (D,N) is an r.b.f. of P;

   2) $D^*D + N^*N = I.$ (4.1)

Completely analogously we define:

DEF.4.1.2. Let $P \in \mathbb{R}^{n*m}(s)$. We say $(\tilde{N},\tilde{D}) \in M(\mathbb{R}H_\infty)$ is a *normalized left–Bezout factorization* (n.l.b.f.) of P if:

   1) $(\tilde{N},\tilde{D})$ is an l.b.f. of P;

   2) $\tilde{N}\tilde{N}^* + \tilde{D}\tilde{D}^* = I.$ (4.2)

Let $\mathbb{R}_p^{n*m}(s)$ denote the set of all n*m matrices with elements in $\mathbb{R}_p(s)$. To prove for each plant $P \in \mathbb{R}_p^{n*m}(s)$ the existence of normalized right– and left–Bezout factorizations, we need the following theorem. It is a generalization of earlier results, stated in [12] and [16].

THEOREM 4.1.1. Let $P \in \mathbb{R}_p^{n*m}(s)$, and suppose [A,B,C,D] is a minimal realization of P. Assume $A \neq [\ ]$. Let X, respectively Y, be the unique positive definite solutions to the Algebraic Riccati Equations:

$$(A-BH^{-1}D^TC)^TX + X(A-BH^{-1}D^TC) - XBH^{-1}B^TX + C^TL^{-1}C = 0 \qquad (4.3)$$

$$(A-BD^TL^{-1}C)Y + Y(A-BD^TL^{-1}C)^T - YC^TL^{-1}CY + BH^{-1}B^T = 0 \qquad (4.4)$$

where $\quad H := I + D^TD,$

$\qquad L := I + DD^T.$

Define $\quad A_c := A - BF$ with $F := H^{-1}(D^TC + B^TX),$

$\qquad A_o := A - KC$ with $K := (BD^T + YC^T)L^{-1}.$

Now define

$$
\begin{bmatrix} -Z_0 & Y_0 \\ \tilde{D}_0 & \tilde{N}_0 \end{bmatrix} := \begin{bmatrix} -H^{1/2}F(sI-A_o)^{-1}K & H^{1/2}[I+F(sI-A_o)^{-1}(B-KD)] \\ L^{-1/2}[I-C(sI-A_o)^{-1}K] & L^{-1/2}[C(sI-A_o)^{-1}(B-KD)+D] \end{bmatrix}
$$

and (4.5)

$$
\begin{bmatrix} -N_0 & \tilde{Y}_0 \\ D_0 & \tilde{Z}_0 \end{bmatrix} := \begin{bmatrix} -[(C-DF)(sI-A_c)^{-1}B+D]H^{-1/2} & [I+(C-DF)(sI-A_c)^{-1}K]L^{1/2} \\ [I-F(sI-A_c)^{-1}B]H^{-1/2} & F(sI-A_c)^{-1}KL^{1/2} \end{bmatrix}
$$

(4.6)

Then (D_0, N_0) and $(\tilde{N}_0, \tilde{D}_0)$ are a normalized right–Bezout and a normalized left–Bezout factorization of P respectively. Moreover we have

$$\begin{bmatrix} -Z_0 & Y_0 \\ \tilde{D}_0 & \tilde{N}_0 \end{bmatrix} \begin{bmatrix} -N_0 & \tilde{Y}_0 \\ D_0 & \tilde{Z}_0 \end{bmatrix} = \begin{bmatrix} I & 0 \\ 0 & I \end{bmatrix}. \tag{4.7}$$

PROOF (i) First we prove that all the matrices defined in (4.5) and (4.6) belong to $M(\mathbb{R}H_\infty)$. To do this, it is sufficient to prove that the matrices A_c and A_o are stable, i.e. all the eigenvalues of the matrices A_c and A_o belong to the LHP.

Suppose X is the unique positive definite solution to the ARE (4.3). Then we have:

$$(A - BH^{-1}D^TC)^TX + X(A - BH^{-1}D^TC) - XBH^{-1}B^TX + C^TL^{-1}C = 0.$$

This equation can be written as

$$(A - BH^{-1}(D^TC + B^TX))^TX + X(A - BH^{-1}(D^TC + B^TX)) + XBH^{-1}B^TX + C^TL^{-1}C = 0.$$

Application of the definition of F yields

$$(A - BF)^TX + X(A - BF) + XBH^{-1}B^TX + C^TL^{-1}C = 0.$$

And because $A - BF = A_c$, we finally get

$$A_c^TX + XA_c + XBH^{-1}B^TX + C^TL^{-1}C = 0. \tag{4.8}$$

Now we need the following auxiliary results:

Claim 1: $C^TL^{-1}C = C^TC - C^TDH^{-1}D^TC$ (4.9)

Proof:

$$C^TC - C^TDH^{-1}D^TC = C^T(I - DH^{-1}D^T)C = C^T(I - D(I + D^TD)^{-1}D^T)C =$$
$$C^T(I - (I + DD^T)^{-1}DD^T)C = C^T((I + DD^T)^{-1}(I + DD^T) - (I + DD^T)^{-1}DD^T)C =$$
$$C^T(I + DD^T)^{-1}C = C^TL^{-1}C.$$

Claim 2: $XBH^{-1}B^TX + C^TL^{-1}C = (C - DF)^T(C - DF) + F^TF$ (4.10)

Proof:

$$(C - DF)^T(C - DF) + F^TF = C^TC - C^TDF - F^TD^TC + F^TD^TDF + F^TF =$$

$$C^TC - C^TDF - F^TD^TC + F^T(I + D^TD)F = C^TC - C^TDF - F^TD^TC + F^THF =$$
$$C^TC - C^TDF - F^TD^TC + F^T(D^TC + B^TX) = C^TC - C^TDF + F^TB^TX =$$
$$C^TC - C^TDH^{-1}(D^TC + B^TX) + (C^TD + XB)H^{-1}B^TX =$$
$$C^TC - C^TDH^{-1}D^TC - C^TDH^{-1}B^TX + C^TDH^{-1}B^TX + XBH^{-1}B^TX =$$
$$C^TL^{-1}C + XBH^{-1}B^TX.$$

From (4.8) and claim 2, we get:

$$A_c^TX + XA_c + (C-DF)^T(C-DF) + F^TF = 0. \qquad (4.11)$$

Now suppose $\lambda \in \sigma(A_c)$. Then there exists a vector $x \neq 0$, such that $A_cx = \lambda x$. Pre-multiplying (4.11) with $\bar{x}^T$ and post-multiplying it with x, we get:

$$2 \operatorname{Re}(\lambda)(\bar{x}^TXx) + \bar{x}^T(C-DF)^T(C-DF)x + \bar{x}^TF^TFx = 0. \qquad (4.12)$$

We now prove that $\bar{x}^T(C-DF)^T(C-DF)x + \bar{x}^TF^TFx > 0$. We know that $(C-DF)^T(C-DF)$ and F^TF are positive semi-definite. Assume that $\bar{x}^T(C-DF)^T(C-DF)x + \bar{x}^TF^TFx = 0$. Then $Fx = 0$ and $Cx = 0$. But $\lambda x = A_cx = (A-BF)x = Ax$, and $Cx = 0$. So λ is an unobservable eigenvalue of A, and this is in contradiction with the assumption that our realization [A,B,C,D] is minimal. So we conclude: $\bar{x}^T(C-DF)^T(C-DF)x + \bar{x}^TF^TFx > 0$.

Now X is positive definite. So $\bar{x}^TXx > 0$, and from (4.12) it follows immediately that $\operatorname{Re}(\lambda) < 0$. So A_c is a stable matrix.

The proof that A_o is a stable matrix is completely analogous. Suppose Y is the unique positive definite solution to the ARE (4.4). Then we have (analogous to (4.8)):

$$A_oY + YA_o^T + YC^TL^{-1}CY + BH^{-1}B^T = 0. \qquad (4.13)$$

Also we have the following facts (analogous to the claims 1 and 2):

Fact 1: $\qquad BH^{-1}B^T = BB^T - BD^TL^{-1}DB^T.$ $\qquad\qquad$ (4.14)

Fact 2: $\qquad BH^{-1}B^T + YC^TL^{-1}CY = (B-KD)(B-KD)^T + KK^T.$ $\qquad$ (4.15)

From (4.13) and (4.15) we get:

$$A_oY + YA_o^T + (B-KD)(B-KD)^T + KK^T = 0. \qquad (4.16)$$

Now suppose $\lambda \in \sigma(A_o)$. Then there exists a row vector y such that $yA_o = \lambda y$. Pre–multiplying (4.16) with y and post–multiplying it with $\bar{y}^T$, we get:

$$2 \, Re(\lambda)(y Y \bar{y}^T) + y(B-KD)(B-KD)^T \bar{y}^T + yKK^T \bar{y}^T = 0. \tag{4.17}$$

Completely analogous to the foregoing case, we can prove that $y(B-KD)(B-KD)^T \bar{y}^T + yKK^T \bar{y}^T > 0$, because otherwise λ would be an uncontrollable eigenvalue of A, which is in contradiction with the assumption that our realization [A,B,C,D] is minimal.

Now Y is positive definite. So $y Y \bar{y}^T > 0$, and from (4.17) it follows immediately that $Re(\lambda) < 0$. So A_o is also a stable matrix.

(ii) Secondly we prove formula (4.7). To do so, we first mention the following fact:

$$KC-BF = -(A-KC) + (A-BF) = A_c - A_o = (sI-A_o) - (sI-A_c) \tag{4.18}$$

We now check the four equalities in (4.7).

a) $Z_0 N_0 + Y_0 D_0 = I$

$$Z_0 N_0 = H^{1/2} F(sI-A_o)^{-1} K[(C-DF)(sI-A_c)^{-1}B+D]H^{-1/2} =$$
$$H^{1/2}\{F(sI-A_o)^{-1}K(C-DF)(sI-A_c)^{-1}B+F(sI-A_o)^{-1}KD\}H^{-1/2}.$$

$$Y_0 D_0 = H^{1/2}[I+F(sI-A_o)^{-1}(B-KD)][I-F(sI-A_c)^{-1}B]H^{-1/2} =$$
$$H^{1/2}\{I-F(sI-A_c)^{-1}B+F(sI-A_o)^{-1}(B-KD)$$
$$-F(sI-A_o)^{-1}(B-KD)F(sI-A_c)^{-1}B\}H^{-1/2}.$$

So, with help of formula (4.18) and the fact that

$$K(C-DF) - (B-KD)F = KC - BF \tag{4.19}$$

we get:

$$Z_0 N_0 + Y_0 D_0 =$$
$$= H^{1/2}\{F(sI-A_o)^{-1}(KC-BF)(sI-A_c)^{-1}B+I-F(sI-A_c)^{-1}B+F(sI-A_o)^{-1}B\}H^{-1/2}$$
$$= H^{1/2}\{F(sI-A_o)^{-1}[(sI-A_o)-(sI-A_c)](sI-A_c)^{-1}B+I+$$
$$-F(sI-A_c)^{-1}B+F(sI-A_o)^{-1}B\}H^{-1/2}$$
$$= H^{1/2}\{F(sI-A_c)^{-1}B-F(sI-A_o)^{-1}B+I-F(sI-A_c)^{-1}B+F(sI-A_o)^{-1}B\}H^{-1/2}$$
$$= H^{1/2}I \, H^{-1/2} = I.$$

b) $-Z_0\tilde{Y}_0 + Y_0\tilde{Z}_0 = 0$

$$-Z_0\tilde{Y}_0 = -H^{1/2}F(sI-A_o)^{-1}K[I+(C-DF)(sI-A_c)^{-1}K]L^{1/2} =$$
$$-H^{1/2}\{F(sI-A_o)^{-1}K+F(sI-A_o)^{-1}K(C-DF)(sI-A_c)^{-1}K\}L^{1/2}.$$

$$Y_0\tilde{Z}_0 = H^{1/2}[I+F(sI-A_o)^{-1}(B-KD)]F(sI-A_c)^{-1}KL^{1/2} =$$
$$H^{1/2}\{F(sI-A_c)^{-1}K+F(sI-A_o)^{-1}(B-KD)F(sI-A_c)^{-1}K\}L^{1/2}.$$

Again, formulae (4.18) and (4.19) give:

$$-Z_0\tilde{Y}_0+Y_0\tilde{Z}_0=$$
$$= H^{1/2}\{F(sI-A_c)^{-1}K-F(sI-A_o)^{-1}K+F(sI-A_o)^{-1}(BF-KC)(sI-A_c)^{-1}K\}L^{1/2} =$$
$$= H^{1/2}\{F(sI-A_c)^{-1}K-F(sI-A_o)^{-1}K+F(sI-A_o)^{-1}[(sI-A_c)-(sI-A_o)](sI-A_c)^{-1}K\}L^{1/2}$$
$$= H^{1/2}\{F(sI-A_c)^{-1}K-F(sI-A_o)^{-1}K+F(sI-A_o)^{-1}K-F(sI-A_c)^{-1}K\}L^{1/2} =$$
$$= H^{1/2}\ 0\ L^{1/2} = 0.$$

c) $-\tilde{D}_0N_0 + \tilde{N}_0D_0 = 0$

$$-\tilde{D}_0N_0 = -L^{-1/2}[I-C(sI-A_o)^{-1}K][(C-DF)(sI-A_c)^{-1}B+D]H^{-1/2} =$$
$$-L^{-1/2}\{(C-DF)(sI-A_c)^{-1}B+D-C(sI-A_o)^{-1}KD+$$
$$-C(sI-A_o)^{-1}K(C-DF)(sI-A_c)^{-1}B\}H^{-1/2}.$$

$$\tilde{N}_0D_0 = L^{-1/2}[C(sI-A_o)^{-1}(B-KD)+D][I-F(sI-A_c)^{-1}B]H^{-1/2} =$$
$$L^{-1/2}\{C(sI-A_o)^{-1}(B-KD)+D-DF(sI-A_c)^{-1}B+$$
$$-C(sI-A_o)^{-1}(B-KD)F(sI-A_c)^{-1}B\}H^{-1/2}.$$

Canceling equal terms and using again (4.18) and (4.19), we get:

$$-\tilde{D}_0N_0+\tilde{N}_0D_0=$$
$$= L^{-1/2}\{C(sI-A_o)^{-1}B-C(sI-A_c)^{-1}B+C(sI-A_o)^{-1}(KC-BF)(sI-A_c)^{-1}B\}H^{-1/2} =$$
$$= L^{-1/2}\{C(sI-A_o)^{-1}B-C(sI-A_c)^{-1}B+$$
$$C(sI-A_o)^{-1}[(sI-A_o)-(sI-A_c)](sI-A_c)^{-1}B\}H^{-1/2} =$$
$$= L^{-1/2}\{C(sI-A_o)^{-1}B-C(sI-A_c)^{-1}B+C(sI-A_c)^{-1}B-C(sI-A_o)^{-1}B\}H^{-1/2} =$$
$$= L^{-1/2}\ 0\ H^{-1/2} = 0.$$

d) $\tilde{D}_0\tilde{Y}_0 + \tilde{N}_0\tilde{Z}_0 = I$

$$\tilde{D}_0\tilde{Y}_0 = L^{-1/2}[I-C(sI-A_o)^{-1}K][I+(C-DF)(sI-A_c)^{-1}K]L^{1/2} =$$
$$L^{-1/2}\{I+(C-DF)(sI-A_c)^{-1}K-C(sI-A_o)^{-1}K+$$
$$-C(sI-A_o)^{-1}K(C-DF)(sI-A_c)^{-1}K\}L^{1/2}.$$

$$\tilde{N}_0\tilde{Z}_0 = L^{-1/2}[C(sI-A_o)^{-1}(B-KD)+D]F(sI-A_c)^{-1}KL^{1/2} =$$
$$L^{-1/2}\{C(sI-A_o)^{-1}(B-KD)F(sI-A_c)^{-1}K+DF(sI-A_c)^{-1}K\}L^{1/2}.$$

Application of (4.18) and (4.19) now gives:

$$\tilde{D}_0\tilde{Y}_0+\tilde{N}_0\tilde{Z}_0=$$
$$= L^{-1/2}\{I+C(sI-A_c)^{-1}K-C(sI-A_o)^{-1}K+C(sI-A_o)^{-1}(BF-KC)(sI-A_c)^{-1}K\}L^{1/2} =$$
$$= L^{-1/2}\{I+C(sI-A_c)^{-1}K-C(sI-A_o)^{-1}K+$$
$$+C(sI-A_o)^{-1}[(sI-A_c)-(sI-A_o)](sI-A_c)^{-1}K\}L^{1/2} =$$
$$= L^{-1/2}\{I+C(sI-A_c)^{-1}K-C(sI-A_o)^{-1}K+C(sI-A_o)^{-1}K-C(sI-A_c)^{-1}K\}L^{1/2} =$$
$$= L^{-1/2} I L^{1/2} = I.$$

   (iii) Now we show that (D_0,N_0) and $(\tilde{N}_0,\tilde{D}_0)$ are an r.b.f. and an l.b.f. of P respectively.

a) First we consider the right–Bezout case. The fact that $(D_0,N_0) \in M(\mathbb{R}H_\infty)$ is already proven in i). So we only have to check the three conditions in definition 1.2.1.

1) Since $D_0 = [I-F(sI-A_c)^{-1}B]H^{-1/2}$, $\det(D_0)$ tends to $\det(H^{-1/2})$ as $s \longrightarrow \infty$. Because $\det(H^{-1/2})$ $= \det((I+D^TD)^{-1/2}) \neq 0$, we know that D_0 is indeed non–singular.
2) The existence of two matrices Y_0 and Z_0 such that $Y_0D_0+Z_0N_0=I$ is already proven in ii.a). From i) we know that these matrices Y_0 and Z_0 belong to $M(\mathbb{R}H_\infty)$.
3) So we only have to prove that $P=N_0D_0^{-1}$. Because D_0 is non–singular it is sufficient to prove that $PD_0=N_0$. To do so, we use the fact that

$$BF = A - (A-BF) = A - A_c = (sI-A_c) - (sI-A). \qquad (4.20)$$

Then we have:

$$PD_0 =$$

$$= [C(sI-A)^{-1}B+D][I-F(sI-A_c)^{-1}B]H^{-1/2} =$$

$$= \{C(sI-A)^{-1}B-C(sI-A)^{-1}BF(sI-A_c)^{-1}B+D-DF(sI-A_c)^{-1}B\}H^{-1/2} =$$

$$= \{C(sI-A)^{-1}B-C(sI-A)^{-1}[(sI-A_c)-(sI-A)](sI-A_c)^{-1}B+D-DF(sI-A_c)^{-1}B\}H^{-1/2} =$$

$$= \{C(sI-A)^{-1}B-C(sI-A_c)^{-1}B+C(sI-A_c)^{-1}B+D-DF(sI-A_c)^{-1}B\}H^{-1/2} =$$

$$= \{(C-DF)(sI-A_c)^{-1}B+D\}H^{-1/2} = N_0.$$

b) The left–Bezout case is completely analogous. In i) we have already proven that $(\tilde{N}_0,\tilde{D}_0) \in$ $M(\mathbb{R}H_\infty)$. Since $\tilde{D}_0 = L^{-1/2}[I-C(sI-A_c)^{-1}K]$, $\det(\tilde{D}_0)$ tends to $\det(L^{-1/2})$ as $s \longrightarrow \infty$, and because $\det(L^{-1/2}) = \det((I+DD^T)^{-1/2}) \neq 0$, $\tilde{D}_0$ is non–singular. The proof of the existence of two matrices $\tilde{Y}_0$ and $\tilde{Z}_0$ in $M(\mathbb{R}H_\infty)$ such that $\tilde{D}_0\tilde{Y}_0+\tilde{N}_0\tilde{Z}_0=I$ is already given in ii.d) and i). So we only have to check the third condition of definition 1.2.2.: $P=\tilde{D}_0^{-1}\tilde{N}_0$. Since $\tilde{D}_0$ is non–singular it is sufficient to prove that $\tilde{D}_0P=\tilde{N}_0$. Now we use the following equality:

$$KC = A - (A-KC) = A - A_o = (sI-A_o) - (sI-A). \qquad (4.21)$$

Then we get:

$$\tilde{D}_0P =$$

$$= L^{-1/2}[I-C(sI-A_o)^{-1}K][C(sI-A)^{-1}B+D] =$$

$$= L^{-1/2}\{C(sI-A)^{-1}B+D-C(sI-A_o)^{-1}KC(sI-A)^{-1}B-C(sI-A_o)^{-1}KD\} =$$

$$= L^{-1/2}\{C(sI-A)^{-1}B+D-C(sI-A_o)^{-1}KD-C(sI-A_o)^{-1}[(sI-A_o)-(sI-A)](sI-A)^{-1}B\} =$$

$$= L^{-1/2}\{C(sI-A)^{-1}B+D-C(sI-A_o)^{-1}KD-C(sI-A)^{-1}B+C(sI-A_o)^{-1}B\} =$$

$$= L^{-1/2}\{C(sI-A_o)^{-1}(B-KD)+D\} = \tilde{N}_0.$$

(iv) Finally we have to prove that both the Bezout factorizations are normalized, i.e. the equalities (4.1) and (4.2) hold. Again we start with the right–Bezout case.

a) We are going to prove that $D_0^*D_0+N_0^*N_0=I$.

$$D_0^*(s)D_0(s) = D_0^T(-s)D_0(s) =$$

$$= H^{-1/2}[I+B^T(sI+A_c^T)^{-1}F^T][I-F(sI-A_c)^{-1}B]H^{-1/2} =$$

$$= H^{-1/2}\{I-F(sI-A_c)^{-1}B+B^T(sI+A_c^T)^{-1}F^T-B^T(sI+A_c^T)^{-1}F^TF(sI-A_c)^{-1}B\}H^{-1/2}.$$

$$(4.22)$$

$N_0^*(s)N_0(s) = \overline{N_\bullet}(-s)N_0(s) =$

$= H^{-1/2}[-B^T(sI+A_c^T)^{-1}(C-DF)^T+D^T][(C-DF)(sI-A_c)^{-1}B+D]H^{-1/2} =$

$= H^{-1/2}\{-B^T(sI+A_c^T)^{-1}(C-DF)^T(C-DF)(sI-A_c)^{-1}B-B^T(sI-A_c^T)^{-1}(C-DF)^TD+$

$\qquad D^T(C-DF)(sI-A_c)^{-1}B+D^TD\}H^{-1/2}.$

(4.23)

Adding (4.22) and (4.23), we get:

$D_0^*(s)D_0(s) + N_0^*(s)N_0(s) =$

$= H^{-1/2}\{-B^T(sI+A_c^T)^{-1}[(C-DF)^T(C-DF)+F^TF](sI-A_c)^{-1}B+(I+D^TD)+$

$\qquad B^T(sI+A_c^T)^{-1}[F^T-(C-DF)^TD]+[D^T(C-DF)-F](sI-A_c)^{-1}B\}H^{-1/2}.$

(4.24)

Now we know that X is the unique positive definite solution to the ARE (4.3), and so equation (4.11) holds. Rearranging (4.11) gives:

$$-(A_c^TX + XA_c) = (C-DF)^T(C-DF) + F^TF.$$

(4.25)

Also we have the following fact:

Claim 3: $F^T - (C-DF)^TD = XB$

(4.26)

Proof:

$F^T-(C-DF)^TD = F^T-C^TD+F^TD^TD = F^T(I+D^TD)-C^TD = F^TH-C^TD =$

$= (C^TD+X^TB)H^{-1}H-C^TD = X^TB = XB.$

Transposing (4.26) and multiplying it with -1, we get:

$$D^T(C-DF) - F = -B^TX.$$

(4.27)

Application of the formulae (4.25), (4.26) and (4.27) in (4.24) finally gives:

$D_0^*(s)D_0(s) + N_0^*(s)N_0(s) =$

$= H^{-1/2}\{B^T(sI+A_c^T)^{-1}[A_c^TX+XA_c](sI-A_c)^{-1}B+H+$

$\qquad B^T(sI+A_c^T)^{-1}XB+-B^TX(sI-A_c)^{-1}B\}H^{-1/2} =$

$= H^{-1/2}\{B^T(sI+A_c^T)^{-1}[(sI+A_c^T)X-X(sI-A_c)](sI-A_c)^{-1}B+H+$

$\qquad B^T(sI+A_c^T)^{-1}XB-B^TX(sI-A_c)^{-1}B\}H^{-1/2} =$

$= H^{-1/2}\{B^TX(sI-A_c)^{-1}B-B^T(sI+A_c^T)^{-1}XB+H+$

$$B^T(sI+A_c^T)^{-1}XB-B^TX(sI-A_c)^{-1}B\}H^{-1/2} =$$
$$= H^{-1/2}\, H\, H^{-1/2} = I.$$

b) The left–Bezout case is again completely analogous:

$$\tilde{N}_0(s)\tilde{N}_0^*(s) = \tilde{N}_0(s)\tilde{N}_0^T(-s) =$$
$$= L^{-1/2}[C(sI-A_o)^{-1}(B-KD)+D][-(B-KD)^T(sI+A_o^T)^{-1}C^T+D^T]L^{-1/2} =$$
$$= L^{-1/2}\{-C(sI-A_o)^{-1}(B-KD)(B-KD)^T(sI+A_o^T)^{-1}C^T+DD^T+$$
$$C(sI-A_o)^{-1}(B-KD)D^T-D(B-KD)^T(sI+A_o^T)^{-1}C^T\}L^{-1/2}. \tag{4.28}$$

$$\tilde{D}_0(s)\tilde{D}_0^*(s) = \tilde{D}_0(s)\tilde{D}_0^T(-s) =$$
$$= L^{-1/2}[I-C(sI-A_o)^{-1}K][I+K^T(sI+A_o^T)^{-1}C^T]L^{-1/2} =$$
$$= L^{-1/2}\{I+K^T(sI+A_o^T)^{-1}C^T-C(sI-A_o)^{-1}K-C(sI-A_o)^{-1}KK^T(sI+A_o^T)^{-1}C^T\}L^{-1/2}. \tag{4.29}$$

Adding (4.28) and (4.29) gives:

$$\tilde{N}_0(s)\tilde{N}_0^*(s) + \tilde{D}_0(s)\tilde{D}_0^*(s) =$$
$$= L^{-1/2}\{-C(sI-A_o)^{-1}[(B-KD)(B-KD)^T+KK^T](sI+A_o^T)^{-1}C^T+(I+DD^T)+$$
$$C(sI-A_o)^{-1}[(B-KD)D^T-K]+[K^T-D(B-KD)^T](sI+A_o^T)^{-1}C^T\}L^{-1/2}. \tag{4.30}$$

Y is the unique positive definite solution to the ARE (4.4), and so (4.16) holds. From (4.16) we immediately get:

$$-(A_oY + YA_o^T) = (B-KD)(B-KD)^T + KK^T. \tag{4.31}$$

Analogous to claim 3, and formula (4.27), we also have the following equalities:

$$(B-KD)D^T - K = -YC^T, \tag{4.32}$$
$$K^T - D(B-KD)^T = CY. \tag{4.33}$$

Using (4.31), (4.32) and (4.33) in (4.30), we find:

$$\tilde{N}_0(s)\tilde{N}_0^*(s) + \tilde{D}_0(s)\tilde{D}_0^*(s) =$$
$$= L^{-1/2}\{C(sI-A_o)^{-1}[A_oY+YA_o^T](sI+A_o^T)^{-1}C^T+L+$$

$$-C(sI-A_o)^{-1}YC^T+CY(sI+A_o^T)^{-1}C^T\}L^{-1/2} =$$

$$= L^{-1/2}\{C(sI-A_o)^{-1}[-(sI-A_o)Y+Y(sI+A_o^T)](sI+A_o^T)^{-1}C^T+L+$$

$$-C(sI-A_o)^{-1}YC^T+CY(sI+A_o^T)^{-1}C^T\}L^{-1/2} =$$

$$= L^{-1/2}\{-CY(sI+A_o^T)^{-1}C^T+C(sI-A_o)^{-1}YC^T+L+$$

$$-C(sI-A_o)^{-1}YC^T+CY(sI+A_o^T)^{-1}C^T\}L^{-1/2} =$$

$$= L^{-1/2} L L^{-1/2} = I.$$

This completes the proof. □

Theorem 4.1.1. gives a method for the construction of an n.r.b.f. and an n.l.b.f. for dynamical systems in $M(\mathbb{R}_p(s))$. For computer purposes this method is implemented in the MATLAB–function ncoprfac. For the details of this implementation, we refer to appendix B.

There is only one little problem left. For non–dynamical systems in $M(\mathbb{R}_p(s))$, theorem 4.1.1. doesn't give a solution to the factorization problem, since for non–dynamical systems it is not possible to find a minimal realization [A,B,C,D] with A ≠ [ ]. But in this case we have the following theorem.

THEOREM 4.1.2. Let $P \in \mathbb{R}_p^{n*m}(s)$ and suppose P = K, with K a *constant* transfermatrix. Define

$$\begin{bmatrix} -Z_0 & Y_0 \\ \tilde{D}_0 & \tilde{N}_0 \end{bmatrix} := \begin{bmatrix} -(I+K^TK)^{-1/2}K^T & (I+K^TK)^{-1/2} \\ (I+KK^T)^{-1/2} & (I+KK^T)^{-1/2}K \end{bmatrix} \tag{4.34}$$

and

$$\begin{bmatrix} -N_0 & \tilde{Y}_0 \\ D_0 & \tilde{Z}_0 \end{bmatrix} := \begin{bmatrix} -K(I+K^TK)^{-1/2} & (I+KK^T)^{-1/2} \\ (I+K^TK)^{-1/2} & K^T(I+KK^T)^{-1/2} \end{bmatrix}. \tag{4.35}$$

Then (D_0,N_0) and $(\tilde{N}_0,\tilde{D}_0)$ are a normalized right–Bezout and a normalized left–Bezout factorization of P respectively. Moreover we have

$$\begin{bmatrix} -Z_0 & Y_0 \\ \tilde{D}_0 & \tilde{N}_0 \end{bmatrix} \begin{bmatrix} -N_0 & \tilde{Y}_0 \\ D_0 & \tilde{Z}_0 \end{bmatrix} = \begin{bmatrix} I & 0 \\ 0 & I \end{bmatrix}. \tag{4.36}$$

PROOF (i) Clearly all the matrices, defined in (4.34) and (4.35) belong to $M(\mathbb{R}H_\infty)$, because they are all constant matrices.

(ii) Secondly we prove formula (4.36). We simply check the four equalities.

a) $Z_0N_0+Y_0D_0 =$

$$= (I+K^TK)^{-1/2}K^TK(I+K^TK)^{-1/2}+(I+K^TK)^{-1/2}(I+K^TK)^{-1/2} =$$

$$= (I+K^TK)^{-1/2}(K^TK+I)(I+K^TK)^{-1/2} = I.$$

b) $-Z_0\tilde{Y}_0+Y_0\tilde{Z}_0 =$

$$= -(I+K^TK)^{-1/2}K^T(I+KK^T)^{-1/2}+(I+K^TK)^{-1/2}K^T(I+KK^T)^{-1/2} = 0.$$

c) $-\tilde{D}_0N_0+\tilde{N}_0D_0 =$

$$= -(I+KK^T)^{-1/2}K(I+K^TK)^{-1/2}+(I+KK^T)^{-1/2}K(I+K^TK)^{-1/2} = 0.$$

d) $\tilde{D}_0\tilde{Y}_0+\tilde{N}_0\tilde{Z}_0 =$

$$= (I+KK^T)^{-1/2}(I+KK^T)^{-1/2}+(I+KK^T)^{-1/2}KK^T(I+KK^T)^{-1/2} =$$

$$= (I+KK^T)^{-1/2}(I+KK^T)(I+KK^T)^{-1/2} = I.$$

(iii) a) First we show that (D_0,N_0) is an r.b.f. of P. We already mentioned that $(D_0,N_0) \in M(\mathbb{R}H_\infty)$. It is also quite clear that D_0 is non–singular, because $D_0 = (I+K^TK)^{-1/2}$ and $(I+K^TK)$ is a positive definite matrix. A solution to the Bezout–identity $Z_0N_0+Y_0D_0=I$ is given in ii.a), so we only have to prove that $P = N_0D_0^{-1}$, and this is quite easy:

$$N_0D_0^{-1} = K(I+K^TK)^{-1/2}(I+K^TK)^{1/2} = K = P.$$

b) The left–Bezout case is completely analogous. Therefore we only give the proof of the third condition of definition 1.2.2.:

$$\tilde{D}_0^{-1}\tilde{N}_0 = (I+KK^T)^{1/2}(I+KK^T)^{-1/2}K = K = P.$$

(iv) Finally we prove that both the Bezout factorizations are normalized. Again we start with the right–Bezout case.

a) $D_0^*D_0+N_0^*N_0 =$

$$= (I+K^TK)^{-1/2}(I+K^TK)^{-1/2}+(I+K^TK)^{-1/2}K^TK(I+K^TK)^{-1/2} =$$

$$= (I+K^TK)^{-1/2}(I+K^TK)(I+K^TK)^{-1/2} = I.$$

b) The left–Bezout case is completely analogous:

$$\tilde{N}_0\tilde{N}_0^*+\tilde{D}_0\tilde{D}_0^* =$$

$$= (I+KK^T)^{-1/2}KK^T(I+KK^T)^{-1/2}+(I+KK^T)^{-1/2}(I+KK^T)^{-1/2} =$$

$$= (I+KK^T)^{-1/2}(KK^T+I)(I+KK^T)^{-1/2} = I.$$

This completes the proof. □

We conclude this section with the following result, which is a generalization of the theorems 4.1.1. and 4.1.2. There we proved by construction the existence of an n.r.b.f. and an n.l.b.f. for a plant $P \in M(\mathbb{R}_p(s))$ in both the dynamical and the non–dynamical case. In [15, p.262] Vidyasagar also proved the existence of n.r.b.f.'s and n.l.b.f.'s for plants $P \in M(\mathbb{R}(s))$. This proof however is not constructive and is therefore not applicable for algorithmization.

THEOREM 4.1.3. Each plant $P \in M(\mathbb{R}(s))$ has a normalized right–Bezout and a normalized left–Bezout factorization.

4.2. Optimally robust control

In the introduction of this chapter, we suggested an approach to find an optimally robust controller. In this section we will describe this method more precisely and show that it really solves the problem of optimally robust control.

Suppose P_0 is the nominal plant, and assume that $C \in S(P_0)$ is a controller that stabilizes P_0. We also assume that the controller C can be implemented exactly, so $C_\lambda = C$. With help of theorem 3.2.5. we can now calculate the stability radius w^{-1} of the controller C, i.e. the radius of the ball around P_0, containing systems P, which are also stabilized by C:

$$ w^{-1} = (\ \| A_0 \| \ \| B_0 \| \ \| (B_0 A_0)^{-1} \| \ )^{-1}. \tag{4.37} $$

So when $\delta(P, P_0) < w^{-1}$, then P is also stabilized by C.

In the problem of optimally robust control, we want to design a compensator C, that not only stabilizes the nominal plant P_0, but also plants in the neighborhood of P_0. We want to make this neighborhood, stabilized by C, as large as possible. Or, to put it more precise, we want to find a controller C with a stability radius as large as possible. Let w_g^{-1} denote this sharpest bound. In this section we will show that a controller C_g that realizes this radius w_g^{-1} is a solution to the problem of optimally robust control. But before we can do this, we first have to give a more formal definition of w_g^{-1}.

Let (D_0, N_0) and $(\tilde{N}_0, \tilde{D}_0)$ be an n.r.b.f. and an n.l.b.f. of P_0 respectively. Let C_0 be one of the stabilizing controllers of P_0 and (Z_0, Y_0) and $(\tilde{Y}_0, \tilde{Z}_0)$ an l.b.f. and an r.b.f. of C_0 respectively, such that

$$ \begin{bmatrix} -Z_0 & Y_0 \\ \tilde{D}_0 & \tilde{N}_0 \end{bmatrix} \begin{bmatrix} -N_0 & \tilde{Y}_0 \\ D_0 & \tilde{Z}_0 \end{bmatrix} = \begin{bmatrix} I & 0 \\ 0 & I \end{bmatrix}. \tag{4.38} $$

Then we know from theorem 1.2.4. that the set of all stabilizing controllers is given by:

$$S(P_0) = \{ \ (Y_0 - R\tilde{N}_0)^{-1}(Z_0 + R\tilde{D}_0) \ | \ R \in M(\mathbb{R}H_\infty), \ |Y_0 - R\tilde{N}_0| \neq 0 \ \}$$
$$= \{ \ (\tilde{Z}_0 + D_0 R)(\tilde{Y}_0 - N_0 R)^{-1} \ | \ R \in M(\mathbb{R}H_\infty), \ |\tilde{Y}_0 - N_0 R| \neq 0 \ \}.$$

(4.39)

From chapter 1, we finally recall formula (1.11):

$$\begin{bmatrix} -Z_0 - R\tilde{D}_0 & Y_0 - R\tilde{N}_0 \\ \tilde{D}_0 & \tilde{N}_0 \end{bmatrix} \begin{bmatrix} -N_0 & \tilde{Y}_0 - N_0 R \\ D_0 & \tilde{Z}_0 + D_0 R \end{bmatrix} = \begin{bmatrix} I & 0 \\ 0 & I \end{bmatrix}.$$

(4.40)

Now let $C = (Y_0 - R\tilde{N}_0)^{-1}(Z_0 + R\tilde{D}_0)$ be one of the controllers that stabilize P_0. The inverse w of the stability radius w^{-1} belonging to C is then given by:

$$w = \| A_0 \| \ \| B \| \ \| (BA_0)^{-1} \| =$$
$$= \left\| \begin{bmatrix} D_0 \\ N_0 \end{bmatrix} \right\| \ \| \ [(Y_0 - R\tilde{N}_0),(Z_0 + R\tilde{D}_0)] \ \| \ \left\| \ ([(Y_0 - R\tilde{N}_0),(Z_0 + R\tilde{D}_0)] \begin{bmatrix} D_0 \\ N_0 \end{bmatrix})^{-1} \right\|.$$

(4.41)

Now $\left\| \begin{bmatrix} D_0 \\ N_0 \end{bmatrix} \right\| = 1$ because (D_0, N_0) is a *normalized* r.b.f. of P_0 and from (4.40) we see that $((Y_0 - R\tilde{N}_0),(Z_0 + R\tilde{D}_0)) \begin{bmatrix} D_0 \\ N_0 \end{bmatrix} = I$, so (4.41) becomes:

$$w = \| \ [(Y_0 - R\tilde{N}_0),(Z_0 + R\tilde{D}_0)] \ \|.$$

(4.42)

With help of (4.42) we can now give a definition of the sharpest stability radius w_g^{-1}. Let $\mathbb{R}H_\infty^{m*n}(s)$ denote the set of all stable m×n transfermatrices with elements in $\mathbb{R}(s)$:

$$\mathbb{R}H_\infty^{m*n}(s) := M(\mathbb{R}H_\infty) \cap \mathbb{R}^{m*n}(s).$$

(4.43)

Then we define

$$w_g := \inf_{R \ \in \ \mathbb{R}H_\infty^{m*n}(s)} \| \ [(Y_0 - R\tilde{N}_0),(Z_0 + R\tilde{D}_0)] \ \|.$$

(4.44)

In section 4.3. we will show that this infimum is really achievable, and how a solution R_g to it, can be calculated. At present we simply assume that this is possible. We define for

$\varepsilon > 0$:

$$K(P_0,\varepsilon) := \{ P \in \mathbb{R}^{n*m}(s) \mid \delta(P_0,P) < \varepsilon \}. \tag{4.45}$$

From theorem 3.2.5. we know that, if R_g is a solution to (4.44), then C_g defined by

$$C_g := (Y_0 - R_g\tilde{N}_0)^{-1}(Z_0 + R_g\tilde{D}_0) \tag{4.46}$$

stabilizes the whole set $K(P_0,w_g^{-1})$. We will now prove that such a controller C_g is a solution to the problem of optimally robust control by showing that its stability radius w_g^{-1} is the sharpest bound in the sense that there are no compensators which can stabilize $K(P_0,\varepsilon)$ with $\varepsilon > w_g^{-1}$. In other words, the largest number ε, such that $K(P_0,\varepsilon)$ can be stabilized by a controller is w_g^{-1}.

First we define an other set, which also describes a neighborhood of the nominal plant P_0. Let $\varepsilon > 0$, then

$$R(P_0,\varepsilon) := \{ P = (N_0 + \Delta_n)(D_0 + \Delta_d)^{-1} \in \mathbb{R}^{n*m}(s) \mid \left\| \begin{bmatrix} \Delta_d \\ \Delta_n \end{bmatrix} \right\| < \varepsilon \}. \tag{4.47}$$

where (D_0,N_0) is an n.r.b.f. of P_0. Remark that $R(P_0,\varepsilon)$ is independent of the choice of the n.r.b.f. of P_0.

The next theorem is a generalization of a result in [17, p.89].

THEOREM 4.2.1. Any controller $C = (Y_0 - R\tilde{N}_0)^{-1}(Z_0 + R\tilde{D}_0)$ stabilizes $R(P_0,\varepsilon)$ if and only if

$$\| [(Y_0 - R\tilde{N}_0),(Z_0 + R\tilde{D}_0)] \| \le \varepsilon^{-1}. \tag{4.48}$$

From this theorem and the definition of w_g (formula (4.44)) we immediately derive:

COROLLARY 4.2.2. The largest number ε such that $R(P_0,\varepsilon)$ can be stabilized by one controller is w_g^{-1}, and if R_g is a solution to (4.44) then

$$C_g = (Y_0 - R_g\tilde{N}_0)^{-1}(Z_0 + R_g\tilde{D}_0)$$

is a controller stabilizing $R(P_0,w_g^{-1})$.

With help of some results, developed in chapter 3, we can now prove the following crucial theorem. It is a generalization of an earlier result in [6], and proved this way by Zhu in [19, pp. 40–41].

THEOREM 4.2.3. If $0 < \varepsilon \leq 1$, then

$$K(P_0,\varepsilon) = R(P_0,\varepsilon). \tag{4.49}$$

PROOF "$\supset$" Let $P = (N_0+\Delta_n)(D_0+\Delta_d)^{-1} \in R(P_0,\varepsilon)$, and $\left\| \begin{bmatrix} \Delta_d \\ \Delta_n \end{bmatrix} \right\| < \varepsilon$. Since (D_0,N_0) is an n.r.b.f. of P_0, we have $\left\| \begin{bmatrix} D_0 \\ N_0 \end{bmatrix} \right\| = 1$, and also $\| (D_0^*,N_0^*) \| = 1$. Because

$$\left\| (D_0^*,N_0^*)\begin{bmatrix} \Delta_d \\ \Delta_n \end{bmatrix} \right\| \leq \| (D_0^*,N_0^*) \| \ \left\| \begin{bmatrix} \Delta_d \\ \Delta_n \end{bmatrix} \right\| < \varepsilon \leq 1$$

we know by lemma 3.2.2. that

$$(D_0^*,N_0^*)\begin{bmatrix} D_0+\Delta_d \\ N_0+\Delta_n \end{bmatrix} = I + (D_0^*,N_0^*)\begin{bmatrix} \Delta_d \\ \Delta_n \end{bmatrix}$$

is a bijective mapping. So, according to lemma 3.1.6., $\delta(P_0,P) < 1$. By lemma 3.1.3. we then know that $\vec{\delta}(P_0,P) = \vec{\delta}(P,P_0) = \delta(P_0,P)$.

Since (D_0,N_0) is an n.r.b.f. of P_0, the following statement holds:

$$\left\| \begin{bmatrix} D_0 \\ N_0 \end{bmatrix}x \right\| = 1 \qquad \text{if and only if} \qquad \| x \| = 1. \tag{4.50}$$

With help of (4.50) we derive:

$$\begin{aligned} \vec{\delta}(P_0,P) \quad &= \sup_{x \in S_{G(P_0)}} \ \inf_{y \in G(P)} \ \| x - y \| = \\ &= \sup_{x \in H_2^m,\ \|x\|=1} \ \inf_{y \in G(P)} \ \left\| \begin{bmatrix} D_0 \\ N_0 \end{bmatrix}x - y \right\|. \end{aligned}$$

And because $\begin{bmatrix} D_0+\Delta_d \\ N_0+\Delta_n \end{bmatrix}H_2^m \subset G(P)$, we find (note that $\begin{bmatrix} D_0+\Delta_d \\ N_0+\Delta_n \end{bmatrix}$ is bounded):

$$\begin{aligned} \vec{\delta}(P_0,P) \quad &\leq \sup_{x \in H_2^m,\ \|x\|=1} \ \inf_{y \in H_2^m} \ \left\| \begin{bmatrix} D_0 \\ N_0 \end{bmatrix}x - \begin{bmatrix} D_0+\Delta_d \\ N_0+\Delta_n \end{bmatrix}y \right\| \leq \\ &\leq \sup_{x \in H_2^m,\ \|x\|=1} \ \left\| \begin{bmatrix} D_0 \\ N_0 \end{bmatrix}x - \begin{bmatrix} D_0+\Delta_d \\ N_0+\Delta_n \end{bmatrix}x \right\| = \end{aligned}$$

$$= \| \begin{bmatrix} \Delta_d \\ \Delta_n \end{bmatrix} \| < \epsilon.$$

So $\delta(P_0,P) = \breve{\delta}(P_0,P) < \epsilon$, and $P \in K(P_0,\epsilon)$.

"⊂" Let $P \in K(P_0,\epsilon)$. Since $\delta(P_0,P) < 1$, we find, according to theorem 3.1.4., that (D,N) given by

$$\begin{bmatrix} D \\ N \end{bmatrix} := \Pi(P) \begin{bmatrix} D_0 \\ N_0 \end{bmatrix}$$

is an r.b.f. of P. And because (D_0,N_0) is an n.r.b.f. of P_0, we have:

$$\| \begin{bmatrix} D \\ N \end{bmatrix} - \begin{bmatrix} D_0 \\ N_0 \end{bmatrix} \| = \| \Pi(P) \begin{bmatrix} D_0 \\ N_0 \end{bmatrix} - \Pi(P_0) \begin{bmatrix} D_0 \\ N_0 \end{bmatrix} \| \leq$$

$$\leq \| \Pi(P) - \Pi(P_0) \| \; \| \begin{bmatrix} D_0 \\ N_0 \end{bmatrix} \| = \| \Pi(P) - \Pi(P_0) \| =$$

$$= \delta(P_0,P) < \epsilon.$$

So $P \in R(P_0,\epsilon)$.

This completes the proof. □

From theorem 4.2.3. and corollary 4.2.2. we finally derive the following solution to the problem of optimally robust control:

COROLLARY 4.2.4. The largest number ϵ such that $K(P_0,\epsilon)$ can be stabilized by one controller is w_g^{-1}. If R_g is a solution to (4.44) then

$$C_g = (Y_0 - R_g \tilde{N}_0)^{-1}(Z_0 + R_g \tilde{D}_0)$$

is a controller stabilizing $K(P_0,w_g^{-1})$.

4.3. Computation of an optimally robust controller and its stability radius

In section 4.2. we reduced the problem of optimally robust control to the computation of an infimum. w_g, and the construction of a matrix R_g, which realizes this bound. In this section we translate this problem to some well known mathematical problems. In the next chapter we will show how these problems can be solved when the plant $P \in M(\mathbb{R}_p(s))$, and we will develop an algorithm that actually computes an optimally robust controller in this case.

At first, however, we start this section presenting the dual versions of the theorems 3.2.5. and 4.2.1. In the original versions of these theorems, we used a (normalized) r.b.f. of the plant P_0 and an l.b.f. of the controller. To derive the dual results, we simply have to change this order: we use a (normalized) l.b.f. of P_0 and an r.b.f. of the controller C. So instead of the first parametrization of all stabilizing controllers in formula (4.39), we take the second one:

$$C = (\tilde{Z}_0 + D_0 R)(\tilde{Y}_0 - N_0 R)^{-1}.$$

Let (D_0, N_0) and $(\tilde{N}_0, \tilde{D}_0)$ be an n.r.b.f. and an n.l.b.f. of the nominal plant $P_0 \in \mathbb{R}^{n*m}(s)$ respectively. Suppose C_0 is one of the stabilizing compensators of P_0, and (Z_0, Y_0) and $(\tilde{Y}_0, \tilde{Z}_0)$ are an l.b.f. and an r.b.f. of C_0 respectively, such that (4.38) holds. Then we have the following dual version of theorem 3.2.5. (Compare this result also with formula (4.42).)

THEOREM 4.3.1. Let $R \in \mathbb{R}H_\infty^{m*n}(s)$ and define

$$C_0 := (\tilde{Z}_0 + D_0 R)(\tilde{Y}_0 - N_0 R)^{-1}. \qquad (4.51)$$

Assume $P_\lambda \in \mathbb{R}^{n*m}(s)$ and $C_\lambda \in \mathbb{R}^{m*n}(s)$ are perturbed versions of P_0 and C_0 respectively. Then the following statement holds:
If

$$\delta(P_\lambda, P_0) + \delta(C_\lambda, C_0) < (\| \begin{bmatrix} \tilde{Y}_0 - N_0 R \\ \tilde{Z}_0 + D_0 R \end{bmatrix} \|)^{-1} \qquad (4.52)$$

then

$$H(P_\lambda, C_\lambda) \text{ is stable.}$$

Analogous to $R(P_0, \varepsilon)$ we can define for $\varepsilon > 0$:

$$T(P_0, \varepsilon) := \{ P = (\tilde{D}_0 + \tilde{\Delta}_d)^{-1}(\tilde{N}_0 + \tilde{\Delta}_n) \in \mathbb{R}^{n*m}(s) \mid \| (\tilde{\Delta}_d, \tilde{\Delta}_n) \| < \varepsilon \}. \qquad (4.53)$$

Now we can formulate the dual version of theorem 4.2.1.

THEOREM 4.3.2. Any controller $C = (\tilde{Z}_0 + D_0 R)(\tilde{Y}_0 - N_0 R)^{-1}$ with $R \in \mathbb{R}H_\infty^{m*n}(s)$ stabilizes $T(P_0, \varepsilon)$ if and only if

$$\left\| \begin{bmatrix} \tilde{Y}_0 - N_0 R \\ \tilde{Z}_0 + D_0 R \end{bmatrix} \right\| \leq \varepsilon^{-1}. \tag{4.54}$$

It is quite clear that the theorems 4.3.1. and 4.3.2. give rise to the following definition of $\tilde{w}_g$, the inverse of the sharpest stability radius in the dual case:

$$\tilde{w}_g := \inf_{R \in \mathbb{R}H_\infty^{m*n}(s)} \left\| \begin{bmatrix} \tilde{Y}_0 - N_0 R \\ \tilde{Z}_0 + D_0 R \end{bmatrix} \right\|. \tag{4.55}$$

Let $\tilde{R}_g$ be a solution to the infimum in (4.55). Then we define the dual optimally robust controller by

$$\tilde{C}_g := (\tilde{Z}_0 + D_0 \tilde{R}_g)(\tilde{Y}_0 - N_0 \tilde{R}_g)^{-1}. \tag{4.56}$$

Now fortunately we have the following theorem, which shows that both the primal and the dual case lead to the same set of solutions to the problem of optimally robust control.

THEOREM 4.3.3. Define

$$V := D_0^* \tilde{Z}_0 - N_0^* \tilde{Y}_0. \tag{4.57a}$$

An alternative formula for V is then given by

$$V = Z_0 \tilde{D}_0^* - Y_0 \tilde{N}_0^*, \tag{4.57b}$$

and for all $R \in \mathbb{R}H_\infty^{m*n}(s)$ we have:

$$\left\| \begin{bmatrix} \tilde{Y}_0 - N_0 R \\ \tilde{Z}_0 + D_0 R \end{bmatrix} \right\| = \left\| [(Y_0 - R\tilde{N}_0), (Z_0 + R\tilde{D}_0)] \right\| = \sqrt{1 + \| V + R \|^2}. \tag{4.58}$$

PROOF First define $Q := \begin{bmatrix} -N_0^* & D_0^* \\ \tilde{D}_0 & \tilde{N}_0 \end{bmatrix}$, then $Q^* = \begin{bmatrix} -N_0 & \tilde{D}_0^* \\ D_0 & \tilde{N}_0^* \end{bmatrix}$. Because (D_0, N_0) and $(\tilde{N}_0, \tilde{D}_0)$ are an n.r.b.f. and an n.l.b.f. of P_0 respectively, Q is a unitary matrix on L_2^{m+n}:

$$QQ^* = \begin{bmatrix} -N_0^* & D_0^* \\ \check{D}_0 & \check{N}_0 \end{bmatrix} \begin{bmatrix} -N_0 & \check{D}_0^* \\ D_0 & \check{N}_0^* \end{bmatrix} = \begin{bmatrix} N_0^*N_0+D_0^*D_0 & -N_0^*\check{D}_0^*+D_0^*\check{N}_0^* \\ -\check{D}_0N_0+\check{N}_0D_0 & \check{D}_0\check{D}_0^*+\check{N}_0\check{N}_0^* \end{bmatrix} = \begin{bmatrix} I & 0 \\ 0 & I \end{bmatrix}.$$

Since multiplication with unitary matrices doesn't change the norms, we get:

$$\left\| \begin{bmatrix} \check{Y}_0-N_0R \\ \check{Z}_0+D_0R \end{bmatrix} \right\| = \left\| Q \begin{bmatrix} \check{Y}_0-N_0R \\ \check{Z}_0+D_0R \end{bmatrix} \right\| = \left\| \begin{bmatrix} D_0^*\check{Z}_0-N_0^*\check{Y}_0 \\ I \end{bmatrix} + \begin{bmatrix} I \\ 0 \end{bmatrix}R \right\|$$

$$(4.59)$$

and because $\begin{bmatrix} 0 & I \\ -I & 0 \end{bmatrix}$ is unitary too, we also have:

$$\| [(Y_0-R\check{N}_0),(Z_0+R\check{D}_0)] \| = \| [(Z_0+R\check{D}_0),(-Y_0+R\check{N}_0)] \| =$$
$$= \| [(Z_0+R\check{D}_0),(-Y_0+R\check{N}_0)]Q^* \| = \| [-I,(Z_0\check{D}_0^*-Y_0\check{N}_0^*)] + R[0,I] \|.$$

$$(4.60)$$

Now we have defined $V := D_0^*\check{Z}_0-N_0^*\check{Y}_0$. We will show that $V = Z_0\check{D}_0^*-Y_0\check{N}_0^*$ too. Because

$$Q \begin{bmatrix} \check{Y}_0 \\ \check{Z}_0 \end{bmatrix} = \begin{bmatrix} -N_0^* & D_0^* \\ \check{D}_0 & \check{N}_0 \end{bmatrix} \begin{bmatrix} \check{Y}_0 \\ \check{Z}_0 \end{bmatrix} = \begin{bmatrix} V \\ I \end{bmatrix},$$

we have

$$\begin{bmatrix} \check{Y}_0 \\ \check{Z}_0 \end{bmatrix} = Q^*Q \begin{bmatrix} \check{Y}_0 \\ \check{Z}_0 \end{bmatrix} = Q^* \begin{bmatrix} V \\ I \end{bmatrix} = \begin{bmatrix} -N_0 & \check{D}_0^* \\ D_0 & \check{N}_0^* \end{bmatrix} \begin{bmatrix} V \\ I \end{bmatrix} = \begin{bmatrix} -N_0 \\ D_0 \end{bmatrix}V + \begin{bmatrix} \check{D}_0^* \\ \check{N}_0^* \end{bmatrix}.$$

And with help of formula (4.38) we find:

$$0 = (-Z_0,Y_0) \begin{bmatrix} \check{Y}_0 \\ \check{Z}_0 \end{bmatrix} = (-Z_0,Y_0) \begin{bmatrix} -N_0 \\ D_0 \end{bmatrix}V + (-Z_0,Y_0) \begin{bmatrix} \check{D}_0^* \\ \check{N}_0^* \end{bmatrix} =$$
$$= V + (-Z_0\check{D}_0^* + Y_0\check{N}_0^*).$$

So it follows that $V = Z_0\check{D}_0^*-Y_0\check{N}_0^*$, and formula (4.57b) is proven.

With help of this result, we can rewrite the formulae (4.59) and (4.60):

$$\left\| \begin{bmatrix} \breve{Y}_0 - N_0 R \\ \breve{Z}_0 + D_0 R \end{bmatrix} \right\| = \left\| \begin{bmatrix} V \\ I \end{bmatrix} + \begin{bmatrix} I \\ 0 \end{bmatrix} R \right\|. \tag{4.61}$$

$$\left\| [(Y_0 - R\tilde{N}_0), (Z_0 + R\tilde{D}_0)] \right\| = \left\| [-I, V] + R[0, I] \right\|. \tag{4.62}$$

We are now in a position to prove (4.58). Let $R \in \mathbb{R}H_\infty^{m*n}(s)$. Then we have:

$$\left\| \begin{bmatrix} V \\ I \end{bmatrix} + \begin{bmatrix} I \\ 0 \end{bmatrix} R \right\|^2 = \left\| \begin{bmatrix} V+R \\ I \end{bmatrix} \right\|^2 = \sup_{x \in H_2^n, \ \|x\|=1} \left\| \begin{bmatrix} V+R \\ I \end{bmatrix} x \right\|^2 =$$

$$= \sup_{x \in H_2^n, \ \|x\|=1} (\|x\|^2 + \| (V+R)x \|^2) =$$

$$= 1 + \sup_{x \in H_2^n, \ \|x\|=1} \| (V+R)x \|^2 = 1 + \| V+R \|^2. \tag{4.63}$$

And also

$$\| [-I, V] + R[0, I] \|^2 = \| [-I, (V+R)] \|^2 =$$

$$= \sup_{x \in H_2^n, \ \|x\|=1} (\|x\|^2 + \| (V+R)x \|^2) =$$

$$= 1 + \sup_{x \in H_2^n, \ \|x\|=1} \| (V+R)x \|^2 = 1 + \| V+R \|^2. \tag{4.64}$$

From (4.61)–(4.64) it follows immediately that for each $R \in \mathbb{R}H_\infty^{m*n}(s)$:

$$\left\| \begin{bmatrix} \breve{Y}_0 - N_0 R \\ \breve{Z}_0 + D_0 R \end{bmatrix} \right\| = \sqrt{1 + \| V+R \|^2} = \| [(Y_0 - R\tilde{N}_0), (Z_0 + R\tilde{D}_0)] \|$$

and this is exactly formula (4.58).

This completes the proof. □

As a direct consequence of theorem 4.3.3. we have that both the characterizations of the sharpest stability radius, w_g^{-1} in the primal case (formula (4.44)) and $\tilde{w}_g^{-1}$ in the dual case (formula (4.55)) are the same, i.e.

$$w_g^{-1} = \tilde{w}_g^{-1}. \tag{4.65}$$

Moreover, if R_g is a solution to the infimum in (4.44) then R_g is also a solution to the infimum in (4.55). And also the other way round: if $\tilde{R}_g$ is a solution to the infimum in (4.55), then $\tilde{R}_g$ is a solution to the infimum in (4.44) too. So by theorem 1.2.5. we know that the optimally robust controller in the primal case, C_g (formula (4.46)) is equal to the optimally robust controller $\tilde{C}_g$ (formula (4.56)) in the dual case. So the primal and dual method deliver the same solutions to the problem of optimally robust control.

Finally, with help of theorem 4.3.3. we are able to actually compute w_g and R_g. Combination of the formulae (4.44) and (4.58) gives:

$$w_g = \inf_{R \in \mathbb{R}H_\infty^{m*n}(s)} \sqrt{1 + \| V+R \|^2} = \sqrt{1 + \inf_{R \in \mathbb{R}H_\infty^{m*n}(s)} \| V+R \|^2} =$$

$$= \sqrt{1 + (\inf_{R \in \mathbb{R}H_\infty^{m*n}(s)} \| V+R \|)^2}. \tag{4.66}$$

So, to determine the stability radius w_g^{-1}, and a matrix R_g which realizes this bound, we have to compute

$$\alpha := \inf_{R \in \mathbb{R}H_\infty^{m*n}(s)} \| V + R \| \tag{4.67}$$

and a solution R_g to this infimum. But this is a well known mathematical problem: a so called standard *Nehari–problem*. And the value of α can be calculated as the norm of a certain *Hankel–operator*. In the next chapter we show how these mathematical problems can be solved when $P \in M(\mathbb{R}_p(s))$. We are then able to develop an algorithm that solves the problem of optimally robust control in this case.

5. AN ALGORITHM TO COMPUTE AN OPTIMALLY ROBUST CONTROLLER AND ITS STABILITY RADIUS

In chapter 4 we translated the problem of optimally robust control to some well known mathematical problems. In this chapter we show how these problems can be solved. Finally we derive an algorithm that computes an optimally robust controller for each plant $P \in M(\mathbb{R}_p(s))$. The stability radius, belonging to this compensator, is also calculated. Note that we confine ourselves to the case $P \in M(\mathbb{R}_p(s))$. This will be a standing assumption during the last chapters of this book.

In the first section of this chapter we treat the non–dynamical case. This case is very simple, and computation of an optimally robust controller can be done by hand. In the rest of the chapter we deal with the dynamical case. First, in the second section, we figure out how a state–space realization of the the the matrix V (formula (4.57)) can be calculated. With this knowledge we are able to compute the value of the infimum α (formula (4.67)) of section 4.3. Here the concept of Hankel–operators turns out to be very useful. In the fourth section we give a solution to the so called Nehari–problem $\inf_{R \in \mathbb{R}H_\infty^{m*n}(s)} \| V+R \|$; we find a matrix R that realizes the value of this infimum. At the end of this chapter, in section 5, we finally put all the pieces together and describe an algorithm that computes an optimally robust controller and its stability radius in the dynamical case.

5.1. The non–dynamical case

As already mentioned in the introduction of this chapter, we treat the non–dynamical case separately. The main reason to do so, was given in section 4.1. When the transfermatrix of a plant $P \in \mathbb{R}_p^{n*m}(s)$ is a constant matrix, we can't use the method of theorem 4.1.1. to construct a normalized r.b.f. and l.b.f. of P. Instead we have to use the construction of theorem 4.1.2. Fortunately this leads to a very simple solution to the problem of optimally robust control.

In section 4.3. we saw that in the end the problem of optimally robust control can be reduced to the computation of the infimum

$$\alpha := \inf_{R \in \mathbb{R}H_\infty^{m*n}(s)} \| V+R \|$$

(where $V = D_0^* Z_0 - N_0^* Y_0 = Z_0 D_0^* - Y_0 N_0^*$) and a solution R to it. We will first solve this problem, and with the solution of it, compute the optimally robust controller and the stability radius belonging to it.

Let $P_0 \in \mathbb{R}_p^{n*m}(s)$ and suppose $P_0 = K$, with K a constant transfermatrix. With help of the formulae (4.34) and (4.35), which describe an n.r.b.f. and an n.l.b.f. of P_0, we can calculate V:

$$V = (I+K^TK)^{-1/2}K^T(I+KK^T)^{-1/2} - (I+K^TK)^{-1/2}K^T(I+KK^T)^{-1/2} = 0.$$

(5.1)

So, we get $\alpha = \inf\limits_{R \in \mathbb{R}H_\infty^{m*n}(s)} \| V+R \| = 0$, and $R_g = 0$ is a matrix which realizes this bound. With help of formula (4.46) and again (4.34) and (4.35) we find the optimally robust controller C_g:

$$C_g = (Y_0-R_g\tilde{N}_0)^{-1}(Z_0+R_g\tilde{D}_0) = Y_0^{-1}Z_0 =$$
$$= (I+K^TK)^{1/2}(I+K^TK)^{-1/2}K^T = K^T.$$

(5.2)

The stability radius w_g^{-1} can be calculated with formula (4.66):

$$w_g^{-1} = (1 + (\inf\limits_{R \in \mathbb{R}H_\infty^{m*n}(s)} \| V+R \|)^2)^{-1/2} = (1 + 0^2)^{-1/2} = 1.$$

(5.3)

So, we have proven the following theorem:

THEOREM 5.1.1. Let $P \in \mathbb{R}_p^{n*m}(s)$ and suppose $P = K$, with K a *constant* transfermatrix. An optimally robust controller C_g stabilizing P is then given by

$$C_g = K^T$$

and the stability radius belonging to C_g is

$$w_g^{-1} = 1.$$

So C_g stabilizes $K(P,1)$.

5.2. A state–space realization of the matrix V

In this section we start with the treatment of the dynamical case. First we have to find a way to compute the matrix $\dot{V}$, or to put it more precisely, we have to construct a state–space realization of the matrix V. It is of course possible to substitute the state–space realizations of theorem 4.1.1. into formula (4.57) and then to construct a state–space realization of V with help of the multiplication and addition formulae given on page ix. But this gives a solution of a very high order. The next theorem shows that there is a much better way to find a state–space realization of V.

THEOREM 5.2.1. Let $P \in \mathbb{R}_p^{n \cdot m}(s)$, and suppose [A,B,C,D] is a minimal realization of P. Assume $A \neq [\ ]$. Let X, respectively Y be the unique positive definite solutions to the Algebraic Riccati Equations (4.3) and (4.4). Define as in theorem 4.1.1.:

$$A_c := A-BF \quad \text{with} \quad F := H^{-1}(D^TC+B^TX) \qquad (\text{where } H = (I+D^TD))$$

$$A_o := A-KC \quad \text{with} \quad K := (BD^T+YC^T)L^{-1} \qquad (\text{where } L = (I+DD^T)).$$

Then we have the following two state–space formulae for the transfermatrix V, defined in (4.57):

$$V = -H^{-1/2}D^TL^{1/2} + H^{-1/2}B^T(sI+A_c^T)^{-1}(I+XY)C^TL^{-1/2} = \qquad (5.4)$$

$$= -H^{1/2}D^TL^{-1/2} + H^{1/2}B^T(I+XY)(sI+A_o^T)^{-1}C^TL^{-1/2}. \qquad (5.5)$$

PROOF With help of the solutions X and Y to the ARE's (4.3) and (4.4), we constructed in theorem 4.1.1. an n.r.b.f. (D_0,N_0) and an n.l.b.f. $(\tilde{N}_0,\tilde{D}_0)$ of P, and matrices $Y_0,Z_0,\tilde{Y}_0$ and $\tilde{Z}_0$ such that (4.7) holds. We here simply substitute these formulae in the definition of V (formula (4.57a)), and calculate a state–space realization of V without using the multiplication and addition formulae of page ix. In this way we get:

$$V = D_0^* \tilde{Z}_0 - N_0^* \tilde{Y}_0$$

with

$$D_0^* \tilde{Z}_0 = D_0^T(-s)\tilde{Z}_0(s) =$$
$$= H^{-1/2}[I+B^T(sI+A_c^T)^{-1}F^T]F(sI-A_c)^{-1}KL^{1/2} =$$
$$= H^{-1/2}\{F(sI-A_c)^{-1}K+B^T(sI+A_c^T)^{-1}F^TF(sI-A_c)^{-1}K\}L^{1/2}, \qquad (5.6)$$

$$-N_0^* \tilde{Y}_0 = -N_0^T(-s)\tilde{Y}_0(s) =$$
$$= H^{-1/2}[B^T(sI+A_c^T)^{-1}(C-DF)^T-D^T][I+(C-DF)(sI-A_c)^{-1}K]L^{1/2} =$$
$$= H^{-1/2}\{B^T(sI+A_c^T)^{-1}(C-DF)^T+B^T(sI+A_c^T)^{-1}(C-DF)^T(C-DF)(sI-A_c)^{-1}K+$$
$$-D^T-D^T(C-DF)(sI-A_c)^{-1}K\}L^{1/2}. \qquad (5.7)$$

Because X is a solution to the ARE (4.3), we know by (4.11) that

$$(C{-}DF)^T(C{-}DF){+}F^TF \; = \; -(A_c^TX{+}XA_c) \; = \; -(sI{+}A_c^T)X{+}X(sI{-}A_c). \tag{5.8}$$

We also recall formula (4.27):

$$F \, - \, D^T(C{-}DF) \; = \; B^TX. \tag{5.9}$$

Application of (5.8) and (5.9) in the addition of (5.6) and (5.7) gives

$$
\begin{aligned}
V \; &= \; D_0^*\tilde{Z}_0 \, - \, N_0^*\tilde{Y}_0 \; = \\
&= \; H^{-1/2}\{B^T(sI{+}A_c^T)^{-1}[(C{-}DF)^T(C{-}DF){+}F^TF](sI{-}A_c)^{-1}K{+} \\
&\qquad\qquad {+}B^T(sI{+}A_c^T)^{-1}(C{-}DF)^T{+}[F{-}D^T(C{-}DF)](sI{-}A_c)^{-1}K{-}D^T\}L^{1/2} \; = \\
&= \; H^{-1/2}\{B^T(sI{+}A_c^T)^{-1}[{-}(sI{+}A_c^T)X{+}X(sI{-}A_c)](sI{-}A_c)^{-1}K{+} \\
&\qquad\qquad {+}B^T(sI{+}A_c^T)^{-1}(C{-}DF)^T{+}B^TX(sI{-}A_c)^{-1}K{-}D^T\}L^{1/2} \; = \\
&= \; H^{-1/2}\{{-}B^TX(sI{-}A_c)^{-1}K{+}B^T(sI{+}A_c^T)^{-1}XK{+}B^T(sI{+}A_c^T)^{-1}(C{-}DF)^T{+} \\
&\qquad\qquad {+}B^TX(sI{-}A_c)^{-1}K{-}D^T\}L^{1/2} \; = \\
&= \; H^{-1/2}\{{-}D^T{+}B^T(sI{+}A_c^T)^{-1}[XK{+}(C{-}DF)^T]\}L^{1/2}. \tag{5.10}
\end{aligned}
$$

Now we need the following auxiliary result:

Claim 1: $[XK{+}(C{-}DF)^T]L^{1/2} = (I{+}XY)C^TL^{-1/2}.$ $\qquad\qquad$ (5.11)

Proof:

$$
\begin{aligned}
[XK{+}(C{-}DF)^T]L^{1/2} \; &= \; \{X(BD^T{+}YC^T)L^{-1}{+}C^T{-}F^TD^T\}L^{1/2} \; = \\
&= \; \{(XBD^T{+}XYC^T)L^{-1}{+}C^T{-}(XB{+}C^TD)H^{-1}D^T\}L^{1/2} \; = \\
&= \; \{(XBD^T{+}XYC^T)L^{-1}{+}C^T{-}(XB{+}C^TD)D^TL^{-1}\}L^{1/2} \; = \\
&= \; \{XBD^TL^{-1}{+}XYC^TL^{-1}{+}C^T{-}XBD^TL^{-1}{-}C^TDD^TL^{-1}\}L^{1/2} \; = \\
&= \; \{XYC^TL^{-1}{+}C^T((I{+}DD^T){-}DD^T)L^{-1}\}L^{1/2} \; = \; (XYC^TL^{-1}{+}C^TL^{-1})L^{1/2} \; = \\
&= \; (I{+}XY)C^TL^{-1/2},
\end{aligned}
$$

where we used the fact that

$$H^{-1}D^T \; = \; (I{+}D^TD)^{-1}D^T \; = \; D^T(I{+}DD^T)^{-1} \; = \; D^TL^{-1}. \tag{5.12}$$

Now by substitution of (5.11) in (5.10), we immediately get:

$$V = -H^{-1/2}D^TL^{1/2} + H^{-1/2}B^T(sI+A_c^T)^{-1}(I+XY)C^TL^{-1/2}$$

and this is exactly formula (5.4).

To prove (5.5), we use the other characterization of V as given in formula (4.57b):

$$V = Z_0\tilde{D}_0^* - Y_0\tilde{N}_0^*.$$

$$
\begin{aligned}
Z_0\tilde{D}_0^* &= Z_0(s)\tilde{D}_0^T(-s) = \\
&= H^{1/2}F(sI-A_o)^{-1}K[I+K^T(sI+A_o^T)^{-1}C^T]L^{-1/2} = \\
&= H^{1/2}\{F(sI-A_o)^{-1}K+F(sI-A_o)^{-1}KK^T(sI+A_o^T)^{-1}C^T\}L^{-1/2}.
\end{aligned}
\tag{5.13}
$$

$$
\begin{aligned}
-Y_0\tilde{N}_0^* &= Y_0(s)\tilde{N}_0^T(-s) = \\
&= H^{1/2}[I+F(sI-A_o)^{-1}(B-KD)][(B-KD)^T(sI+A_o^T)^{-1}C^T-D^T]L^{-1/2} = \\
&= H^{1/2}\{F(sI-A_o)^{-1}(B-KD)(B-KD)^T(sI+A_o^T)^{-1}C^T-D^T+ \\
&\qquad +(B-KD)^T(sI+A_o^T)^{-1}C^T-F(sI-A_o)^{-1}(B-KD)D^T\}L^{-1/2}.
\end{aligned}
\tag{5.14}
$$

Since Y is a solution to the ARE (4.4), we know, according to (4.16):

$$(B-KD)(B-KD)^T+KK^T = -A_oY-YA_o^T = (sI-A_o)Y-Y(sI+A_o^T).\tag{5.15}$$

We also recall formula (4.32)

$$K - (B-KD)D^T = YC^T.\tag{5.16}$$

With help of (5.15) and (5.16), addition of (5.13) and (5.14) yields

$$
\begin{aligned}
V &= Z_0\tilde{D}_0^* - Y_0\tilde{N}_0^* = \\
&= H^{1/2}\{F(sI-A_o)^{-1}[(B-KD)(B-KD)^T+KK^T](sI+A_o^T)^{-1}C^T-D^T+ \\
&\qquad +F(sI-A_o)^{-1}[K-(B-KD)D^T]+(B-KD)^T(sI+A_o^T)^{-1}C^T\}L^{-1/2} = \\
&= H^{1/2}\{F(sI-A_o)^{-1}[(sI-A_o)Y-Y(sI+A_o^T)](sI+A_o^T)^{-1}C^T-D^T+ \\
&\qquad +F(sI-A_o)^{-1}YC^T+(B-KD)^T(sI+A_o^T)^{-1}C^T\}L^{-1/2} = \\
&= H^{1/2}\{FY(sI+A_o^T)^{-1}C^T-F(sI-A_o)^{-1}YC^T-D^T+ \\
&\qquad +F(sI-A_o)^{-1}YC^T+(B-KD)^T(sI+A_o^T)^{-1}C^T\}L^{-1/2} =
\end{aligned}
$$

$$= H^{1/2}\{-D^T+[FY+(B-KD)^T](sI+A_o^T)^{-1}C^T\}L^{-1/2}. \tag{5.17}$$

Finally we have to use the following result:

Claim 2: $H^{1/2}[FY+(B-KD)^T] = H^{-1/2}B^T(I+XY)$ (5.18)

Proof:

With help of formula (5.12) we find:

$H^{1/2}[FY+(B-KD)^T] = H^{1/2}\{H^{-1}(D^TC+B^TX)Y+B^T-D^TK^T\} =$

$= H^{1/2}\{H^{-1}D^TCY+H^{-1}B^TXY+B^T-D^TL^{-1}(DB^T+CY^T)\} =$

$= H^{1/2}\{H^{-1}D^TCY+H^{-1}B^TXY+B^T-H^{-1}D^TDB^T-H^{-1}D^TCY\} =$

$= H^{1/2}\{H^{-1}B^TXY+(H^{-1}(I+D^TD)-H^{-1}D^TD)B^T\} = H^{1/2}\{H^{-1}B^TXY+H^{-1}B^T\} =$

$= H^{-1/2}B^T(I+XY).$

Substituting (5.18) into (5.17), we get:

$$V = -H^{1/2}D^TL^{-1/2} + H^{-1/2}B^T(I+XY)(sI+A_o^T)^{-1}C^TL^{-1/2}$$

and this is exactly formula (5.5).

This completes the proof, because in theorem 4.3.3. the equality of the two formulae for V was already shown. □

Although not necessary, it is also possible to prove the equivalence of (5.4) and (5.5) directly. First of all application of (5.12) yields:

$$H^{-1/2}D^TL^{1/2} = H^{-1/2}HH^{-1}D^TL^{1/2} = H^{1/2}D^TL^{-1}L^{1/2} = H^{1/2}D^TL^{-1/2}.$$

So to prove the equality of (5.4) and (5.5) it is necessary and sufficient to prove that:

$$(sI+A_c^T)^{-1}(I+XY) = (I+XY)(sI+A_o^T)^{-1}. \tag{5.19}$$

But this equality can also be simplified:

$$(sI+A_c^T)^{-1}(I+XY) = (I+XY)(sI+A_o^T)^{-1}$$

$\Leftrightarrow$

$$(I+XY)(sI+A_o^T) = (sI+A_c^T)(I+XY)$$

$\Leftrightarrow$

$$(I+XY)A_o^T = A_c^T(I+XY). \tag{5.20}$$

Now we have:

$$(I+XY)A_o^T = (I+XY)(A^T-C^TK^T) = (I+XY)(A^T-C^TL^{-1}(DB^T+CY)) =$$
$$= A^T-C^TL^{-1}DB^T-C^TL^{-1}CY+XYA^T-XYC^TL^{-1}DB^T-XYC^TL^{-1}CY \tag{5.21}$$

and

$$A_c^T(I+XY) = (A^T-F^TB^T)(I+XY) = (A^T-(C^TD+XB)H^{-1}B^T)(I+XY) =$$
$$= A^T-C^TDH^{-1}B^T-XBH^{-1}B^T+A^TXY-C^TDH^{-1}B^TXY-XBH^{-1}B^TXY. \tag{5.22}$$

Since $DH^{-1} = L^{-1}D$, combination of (5.20), (5.21) and (5.22) gives:

$$(I+XY)A_o^T = A_c^T(I+XY)$$

$\Leftrightarrow$

$$-C^TL^{-1}CY+XYA^T-XYC^TL^{-1}DB^T-XYC^TL^{-1}CY =$$
$$= -XBH^{-1}B^T+A^TXY-C^TDH^{-1}B^TXY-XBH^{-1}B^TXY$$

$\Leftrightarrow$

$$XY(A^T-C^TL^{-1}DB^T)-XYC^TL^{-1}CY+XBH^{-1}B^T =$$
$$= (A^T-C^TDH^{-1}B^T)XY-XBH^{-1}B^TXY+C^TL^{-1}CY$$

$\Leftrightarrow$

$$X\{Y(A^T-C^TL^{-1}DB^T)-YC^TL^{-1}CY+BH^{-1}B^T\} =$$
$$= \{(A^T-C^TDH^{-1}B^T)X-XBH^{-1}B^TX+C^TL^{-1}C\}Y \tag{5.23}$$

With help of the ARE's (4.3) and (4.4) we can transform (5.23) to:

$$(I+XY)A_o^T = A_c^T(I+XY)$$

$\Leftrightarrow$

$$X(-(A-BD^TL^{-1}C)Y) = (-X(A-BH^{-1}D^TC))Y.$$

And using the fact that $D^TL^{-1} = H^{-1}D^T$ (formula (5.12)), this last equation is easy to prove:

$$X(-(A-BD^TL^{-1}C)Y) = -XAY+XBD^TL^{-1}CY = -XAY+XBH^{-1}D^TCY =$$

$$= (-X(A-BH^{-1}D^TC))Y.$$

So this completes the proof of formula (5.20) and therefore also the direct proof of the equality of (5.4) and (5.5).

We remark here that the state–space realizations of V given in (5.4) and (5.5) have the same order as the original plant P. But this is not the only property of these formulae. If [A,B,C,D] is a minimal realization of P, the given state–space formulae of V are also minimal. This result is stated in the next theorem.

THEOREM 5.2.2. Let $P \in \mathbb{R}_p^{n*m}(s)$ and suppose [A,B,C,D] is a *minimal realization* of P. Assume $A \neq [\ ]$. Then the state–space realizations of the matrix V, associated with P as defined in (4.57), and given by the formulae (5.4) and (5.5):

$$V(s) = [-A_c^T, (I+XY)C^T L^{-1/2}, H^{-1/2} B^T, -H^{-1/2} D^T L^{1/2}] \qquad (5.24a)$$

$$= [-A_o^T, C^T L^{-1/2}, H^{-1/2} B^T (I+XY), -H^{-1/2} D^T L^{-1/2}] \qquad (5.24b)$$

are also *minimal*.

PROOF a) We start with the proof of the minimality of the realization (5.24a).

i) First we show that the pair $(-A_c^T, (I+XY)C^T L^{-1/2})$ is controllable. Because $L^{-1/2}$ is invertible it is enough to prove that $(A_c^T, (I+XY)C^T)$ is controllable. After transposition this is equivalent with the claim that $(C(I+YX),A_c)$ is observable. And this last statement is easy to prove with help of formula (5.20). Because X and Y are the solutions of the ARE's (4.3) and (4.4) we have (5.20):

$$(I+XY)A_o^T = A_c^T(I+XY).$$

From this we immediately get (after transposition):

$$A_o(I+YX) = (I+YX)A_c. \qquad (5.25)$$

Since (I+YX) is positive definite and therefore invertible, we have

$$A_c = (I+YX)^{-1}A_o(I+YX). \qquad (5.26)$$

With (5.26) we derive that $(C(I+YX),A_c)$ is observable if and only if $(C(I+YX),(I+YX)^{-1}A_o(I+YX))$ is observable. Because we can consider (I+YX) as a basis transformation, this is equivalent with (C, A_o) is observable. So, since $A_o = A - KC$, we have to prove that $(C, A - KC)$ is observable. But because $[A, B, C, D]$ is a minimal realization of P, this last statement is immediately clear from the fact that (C, A) is observable.

ii) The second part of the proof is quite easy. $(H^{-1/2}B^T, -A_c^T)$ is observable if and only if $(A_c, BH^{-1/2})$ is controllable. Since $H^{-1/2}$ is invertible this is equivalent with (A_c, B) is controllable. And this last statement follows immediately from the definition $A_c = A - BF$ and the fact that (A, B) is controllable. So $(A - BF, B)$ is clearly controllable.

b) The proof of the minimality of the realization (5.24b) of V is completely analogous, and is therefore omitted. □

For computer purposes the construction of this state–space realization of V has been implemented in the MATLAB–function speccopr. More information can be found in appendix B.

To conclude this section we finally remark that, apart from the constant term $-H^{-1/2}D^T L^{1/2}$, the matrix V is completely anti–stable. In theorem 4.1.1. we proved that the matrices A_c and A_o are stable matrices, so $-A_c^T$ and $-A_o^T$ are anti–stable: all their eigenvalues belong to the RHP. From this fact we derive that in the dynamical case the stability radius w_g^{-1} is always smaller than one. Because V is anti–stable we know that

$$\alpha = \inf_{R \in \mathbb{R}H_\infty^{m*n}(s)} \| V + R \| > 0.$$

And with help of this inequality we get

$$w_g = \sqrt{1 + (\inf_{R \in \mathbb{R}H_\infty^{m*n}(s)} \| V + R \|)^2} > 1,$$

so

$$w_g^{-1} = 1/w_g < 1. \tag{5.27}$$

This result is not very surprising. Since we know that the gap between two plants is always smaller or equal to one, we can't expect to find an optimally robust controller with a stability radius equal to one in the dynamical case. The exact calculation of the stability radius w_g^{-1} however, will be the subject of the next section.

5.3. Computation of the optimal stability radius w_g^{-1}; Hankel–operators

With help of the state–space realization of the matrix V, found in the last section, we are now able to derive a method to calculate the sharpest stability radius w_g^{-1} for each plant $P \in M(\mathbb{R}_p(s))$, as defined in (4.44). The infimum we want to calculate, turns out to be equal to the norm of a certain *Hankel–operator*. First of all, however, we have to give a few more definitions. (In this section we often drop the superscripts for notational convenience. The dimensions are then clear from the context.)

DEF.5.3.1. The *space* L_2^n consists of all functions $x : i\mathbb{R} \longrightarrow \mathbb{C}^n$, which are square–(Lebesgue) integrable. L_2^n is a Hilbert–space under the inner–product

$$(x,y) := (2\pi)^{-1} \int_{-\infty}^{\infty} x(i\omega)^* y(i\omega) d\omega. \tag{5.28}$$

We see that H_2 is a closed subspace of L_2. Of course its orthogonal complement $H_2^\perp$ is closed too. So it is possible to introduce the following orthogonal projections: Π_1, the orthogonal projection from L_2 onto $H_2^\perp$, and Π_2, the orthogonal projection from L_2 onto H_2.

DEF.5.3.2. The *space* L_∞ consists of all functions $F : i\mathbb{R} \longrightarrow \mathbb{C}$ for which $| F(i\omega) |$ is essentially bounded (bounded except possibly on a set of measure zero). The L_∞–norm of F is defined as

$$\| F \|_\infty := \underset{\omega \in \mathbb{R}}{ess\ sup} \ | F(i\omega) |. \tag{5.29}$$

With this norm, L_∞ is a Banach–space.

DEF.5.3.3. $\mathbb{R}L_\infty := \mathbb{R}(s) \cap L_\infty$.

So, $\mathbb{R}L_\infty$ is the subspace of L_∞ which consists of all real–rational proper functions without poles on the imaginary axis. Let $M(L_\infty)$ and $M(\mathbb{R}L_\infty)$ denote the sets of all matrices with elements in L_∞ and $\mathbb{R}L_\infty$ respectively. A matrix $F \in M(L_\infty)$ can then be seen as an operator mapping L_2 into L_2: $FL_2 \subset L_2$. We can now give the following definition:

DEF.5.3.4. Let $F \in M(L_\infty)$. The *Laurent–operator with symbol* F, denoted Λ_F, is an operator from L_2 to L_2 and is defined as

$$\Lambda_F g := Fg \tag{5.30}$$

$\| \Lambda_F \| = \| F \|_\infty$, so Λ_F is bounded.

We are now ready to introduce the Hankel–operator with symbol F.

DEF.5.3.5. Let $F \in M(L_\infty)$. The *Hankel-operator with symbol* F, denoted Γ_F, maps H_2 to $H_2^\perp$, and is defined as

$$\Gamma_F := \Pi_1 \Lambda_F \mid_{H_2}. \tag{5.31}$$

Remark that when $F \in M(H_\infty)$, then $\Gamma_F = 0$ because $FH_2 \subset H_2$, so the projection onto $H_2^\perp$, Π_1, maps this whole set to zero.

The relation between Hankel-operators and the infimum we want to calculate is given by Nehari's theorem:

THEOREM 5.3.1. Let $V \in M(RL_\infty)$. Then we have:

i) $\displaystyle \inf_{R \in M(RH_\infty)} \| V-R \|_\infty = \| \Gamma_V \|$ (5.32)

ii) There exists a matrix $R \in M(RH_\infty)$ such that $\| V-R \| = \| \Gamma_V \|$.

PROOF This theorem was stated in a slightly different way by Francis in [4, p.59]. For the proof however, we refer to [13]. How to find an R such that $\| V-R \| = \| \Gamma_V \|$ will be the subject of the next section.

With this theorem we are able to rewrite the formula for w_g, the inverse of the maximal stability radius.

$$w_g = \sqrt{1 + (\inf_{R \in RH_\infty^{m*n}(s)} \| V+R \|)^2} = \sqrt{1 + \| \Gamma_V \|^2}. \tag{5.33}$$

So, to compute the maximal stability radius we only have to find a method to calculate the norm of the Hankel-operator with symbol V, when a state-space realization of V is given. In the rest of this section we describe such a method. For a complete proof of it, and more details, we refer to [4, ch.5].

Let $V \in M(RL_\infty)$. There exists a unique decomposition (for example by partial-fraction decomposition) $V = V_1 + V_2$, where V_1 is strictly proper and analytic in $Re(s) \leq 0$, and V_2 is proper and analytic in $Re(s) \geq 0$, i.e. $V_2 \in M(RH_\infty)$. So $\Gamma_V = \Gamma_{V_1} + \Gamma_{V_2} = \Gamma_{V_1}$ because $\Gamma_{V_2} = 0$. Introduce a (minimal) realization of V_1:

$$V_1(s) = [A,B,C,0]. \tag{5.34}$$

Then the matrix A is anti-stable. In the case we are interested in, with V as in formula

(5.4), this decomposition is quite simple. V_2 consists only of the constant matrix $-H^{-1/2}D^T L^{1/2}$ and a realization of V_1 is given by

$$V_1(s) = [-A_c^T, (I+XY)C^T L^{-1/2}, H^{-1/2}B^T, 0]. \tag{5.35}$$

We now return to the general case, with V_1 defined as in (5.34) and A anti–stable. We define two auxiliary operators, the *controllability operator* ψ_c:

$$\psi_c : L_2[0,\infty) \longrightarrow \mathbb{C}^n : \quad \psi_c u := \int_0^\infty e^{-A\tau} B u(\tau) d\tau \tag{5.36}$$

and the *observability operator* ψ_o:

$$\psi_o : \mathbb{C}^n \longrightarrow L_2(-\infty, 0] : \quad (\psi_o x)(t) := Ce^{At}x \qquad (t < 0). \tag{5.37}$$

By the inverse Laplace–transformation we can calculate the time–domain analogue Γ_{v_1} of the Hankel–operator Γ_{V_1}. This operator turns out to be equal to:

$$\Gamma_{v_1} = \psi_o \psi_c. \tag{5.38}$$

(For a proof, see [4, p.54].) Since the Laplace–transformation is an isometry, it preserves the norms of the operators. From this fact the equality $\| \Gamma_{V_1} \| = \| \Gamma_{v_1} \|$ follows immediately.

We also need a little more operator theory. Let Φ be an operator from a Hilbert–space X to a Hilbert–space Y. Let Φ^* denote the adjoint of Φ. Then $\Phi^*\Phi$ is self–adjoint and we have

$$\| \Phi \|^2 = \| \Phi^*\Phi \| \tag{5.39}$$

(for a proof, see [11, p.199]). From this equation we immediately derive that $\| \Gamma_{V_1} \| = \sqrt{\| \Gamma_{V_1}^* \Gamma_{V_1} \|}$. So, instead of computing $\| \Gamma_{V_1} \|$, we can also calculate $\| \Gamma_{V_1}^* \Gamma_{V_1} \|$, and this can be done with help of the next theorem.

THEOREM 5.3.2. The spectrum $\sigma(\Gamma_{V_1}^* \Gamma_{V_1})$ of $\Gamma_{V_1}^* \Gamma_{V_1}$ is not empty. The eigenvalues of $\Gamma_{V_1}^* \Gamma_{V_1}$ are real and nonnegative and the largest of them equals $\| \Gamma_{V_1}^* \Gamma_{V_1} \|$.

PROOF A proof of the first claim is given in [4, p.57]. The rest of the proof can be found in [11, sec.9.1–9.3].

We now define the *controllability* and *observability gramians*:

$$L_c := \int_0^\infty e^{-At}BB^Te^{-A^Tt}dt \tag{5.40}$$

$$L_o := \int_0^\infty e^{-A^Tt}C^TCe^{-At}dt. \tag{5.41}$$

It is not difficult to show that L_c and L_o are the unique solutions of the *Lyapunov equations*:

$$AL_c + L_cA^T = BB^T \tag{5.42}$$

$$A^TL_o + L_oA = C^TC. \tag{5.43}$$

It is also quite easy to prove that the matrix representations of $\psi_c\psi_c^*$ and $\psi_o^*\psi_o$ are L_c and L_o respectively. This last fact gives rise to the following theorem.

THEOREM 5.3.3. The operator $\Gamma_{V_1}^*\Gamma_{V_1}$ and the matrix L_cL_o have the same nonzero eigenvalues.

PROOF i) Let $\lambda \neq 0$ be an eigenvalue of $\Gamma_{V_1}^*\Gamma_{V_1}$. Then obviously λ is an eigenvalue of $\Gamma_{v_1}^*\Gamma_{v_1}$, the time–domain analogue of $\Gamma_{V_1}^*\Gamma_{V_1}$, which equals $\psi_c^*\psi_o^*\psi_o\psi_c$. So there exists a $u \neq 0$ in $L_2[0,\infty)$ such that

$$\psi_c^*\psi_o^*\psi_o\psi_cu = \lambda u. \tag{5.44}$$

Pre–multiplying (5.44) by ψ_c and defining $x := \psi_cu$, we get:

$$L_cL_ox = \psi_c\psi_c^*\psi_o^*\psi_o\psi_cu = \psi_c\lambda u = \lambda\psi_cu = \lambda x.$$

Clearly $x \neq 0$, because otherwise ψ_cu were to equal zero, and so would λu from (5.44). Since both λ and u are nonzero, this is not possible and λ is an eigenvalue of L_cL_o.

ii) Let $\lambda \neq 0$ be an eigenvalue of L_cL_o, and $x \neq 0$ a corresponding eigenvector, so

$$L_cL_ox = \lambda x. \tag{5.45}$$

Pre–multiplying (5.45) by $\psi_c^*L_o$ and defining $u := \psi_c^*L_ox$, we derive:

$$\psi_c^*\psi_o^*\psi_o\psi_cu = \psi_c^*\psi_o^*\psi_o\psi_c\psi_c^*L_ox = \psi_c^*L_oL_cL_ox = \psi_c^*L_o\lambda x = \lambda\psi_c^*L_ox = \lambda u.$$

From (5.45) it also follows that $\lambda x = L_c L_o x = \psi_c \psi_c^* L_o x = \psi_c u$. Because λ and x are both nonzero, $\psi_c u \neq 0$, and this is only possible when $u \neq 0$. Therefore λ is an eigenvalue of $\psi_c^* \psi_o^* \psi_o \psi_c$, and hence of $\Gamma_{V_1}^* \Gamma_{V_1}$. $\square$

Combination of theorem 5.3.2. and 5.3.3. yields:

COROLLARY 5.3.4. $\| \Gamma_{V_1}^* \Gamma_{V_1} \|$ equals the largest eigenvalue of $L_c L_o$.

So we have found a way to calculate $\| \Gamma_{V_1} \| = \sqrt{\| \Gamma_{V_1}^* \Gamma_{V_1} \|}$.

To conclude this section, we summarize the results of it in the form of an algorithm, which computes the norm of the Hankel–operator with symbol V.

Step 1: Find a realization [A,B,C,0] of the strictly proper and anti–stable part of V(s):

$$V(s) = [A,B,C,0] + \text{(a matrix in } M(\mathbb{R}H_\infty)).$$

(In our particular case this factorization is given by

$$V(s) = [-A_c^T, (I+XY)C^T L^{-1/2}, H^{-1/2} B^T, 0] + (-H^{-1/2} D^T L^{1/2})$$

in accordance with formula (5.4).)

Step 2: Solve the Lyapunov equations (5.42) and (5.43) for L_c and L_o.

Step 3: Then $\| \Gamma_V \| = \sqrt{\lambda_{max}(L_c L_o)}$, where $\lambda_{max}(L_c L_o)$ is the largest eigenvalue of $L_c L_o$.

To calculate the maximal stability radius, we only have to take one extra step:

Step 4: $w_g^{-1} = (1 + \| \Gamma_V \|^2)^{-1/2} = (1 + \lambda_{max}(L_c L_o))^{-1/2}$. (5.46)

The algorithm described above (without step 4), is implemented in the MATLAB–function hankel. For the details of this implementation we refer to appendix C.

5.4. Solving the Nehari–problem

In the last section we derived a method to compute w_g^{-1}, the maximal stability radius that can be achieved by one compensator. To do so, we had to calculate the value α of the infimum

$$\alpha = \inf_{R \ \in \ \mathbb{R}H_\infty^{m*n}(s)} \| \ V+R \ \| = \| \ \Gamma_V \ \|.$$

But we are not only interested in the stability radius itself, but also in the compensator C that realizes this bound. From corollary 4.2.4. we know that for the construction of such a compensator, we need a matrix R such that $\| \ V+R \ \| = \| \ \Gamma_V \ \|$. Nehari's theorem (theorem 5.3.1.) assures that such a matrix in fact exists. The problem to find such an R is called a *Nehari–problem*.

In this section we will describe a method to solve the Nehari–problem in the sub–optimal case. I.e. for each $\varepsilon > 0$, we give a construction of a matrix $R_\varepsilon \in \mathbb{R}H_\infty^{m*n}(s)$ such that

$$\| \ V+R_\varepsilon \ \| \leq \| \ \Gamma_V \ \| \ (1+\varepsilon). \tag{5.47}$$

However, it is also possible to construct an exact solution to the Nehari–problem. In section 6.2. we will return to this subject.

At first we replace the plus–sign in the infimum by a minus–sign. I.e. for each $\varepsilon > 0$ we search for a matrix $R_\varepsilon \in M(\mathbb{R}H_\infty)$ such that $\| \ V-R_\varepsilon \ \| \leq \| \ \Gamma_V \ \| \ (1+\varepsilon)$. In this way we get a standard Nehari–problem as described in most literature. So in the end we have to replace R_ε by $-R_\varepsilon$ to solve the original problem.

Analogous as in the last section it is possible to decompose V in a stable and a strictly proper anti–stable part:

$$V(s) = V_1(s) + V_2(s) = [A,B,C,0] + (a \ matrix \ in \ M(\mathbb{R}H_\infty))$$

with A anti–stable. The matrix $V_2(s)$ in $M(\mathbb{R}H_\infty)$ can be approximated exactly in $M(\mathbb{R}H_\infty)$ (by itself), so in this way we can reduce the sub–optimal Nehari–problem to the following problem: "Let V_1 be completely anti–stable, i.e. V_1 is strictly proper and $V_1^* \in M(\mathbb{R}H_\infty)$. Find for each $\varepsilon > 0$ a matrix $R_\varepsilon \in M(\mathbb{R}H_\infty)$ such that $\| \ V_1-R_\varepsilon \ \| \leq \| \ \Gamma_{V_1} \ \| \ (1+\varepsilon)$." The original problem can then be solved by adding the matrix V_2 to our solution R_ε.

We will now present a solution to the Nehari–problem as stated in [4, sec.8.3.]. A proof of it would take us too far afield. We therefore omit it here, and refer for it to [4].

Let $V_1(s) = [A,B,C,0]$ with A anti-stable (so $V_1^* \in M(\mathbb{R}H_\infty)$), and suppose $\| \Gamma_{V_1} \| < 1$. Let L_c and L_o be the unique solutions to the Lyapunov equations

$$AL_c + L_cA^T = BB^T \tag{5.48}$$

$$A^TL_o + L_oA = C^TC. \tag{5.49}$$

Define $N := (I-L_oL_c)^{-1}$, and the matrix L by

$$L := \begin{bmatrix} L_1 & L_2 \\ L_3 & L_4 \end{bmatrix} \tag{5.50}$$

with

$$L_1(s) = [A,-L_cNC^T,C,I] \tag{5.51a}$$

$$L_2(s) = [A,N^TB,C,0] \tag{5.51b}$$

$$L_3(s) = [-A^T,NC^T,-B^T,0] \tag{5.51c}$$

$$L_4(s) = [-A^T,NL_oB,B^T,I]. \tag{5.51d}$$

Then we have the following theorem, which shows that there is not a unique solution to the Nehari-problem, but a whole set of solutions.

THEOREM 5.4.1. Let $V_1 \in M(\mathbb{R}L_\infty)$, strictly proper, and suppose $\| \Gamma_{V_1} \| < 1$, and $V_1^* \in M(\mathbb{R}H_\infty)$. The set of all matrices $R \in M(\mathbb{R}H_\infty)$ such that $V_1-R \in M(\mathbb{R}H_\infty)$ and $\| V_1-R \| \leq 1$ is given by:

$$R = V_1 - X_1X_2^{-1}, \tag{5.52}$$

with $\begin{bmatrix} X_1 \\ X_2 \end{bmatrix} = L \begin{bmatrix} Y \\ I \end{bmatrix},$ (5.53)

where $Y \in M(\mathbb{R}H_\infty)$, $\| Y \|_\infty \leq 1$.

With help of this theorem we can now solve the Nehari-problem in the sub-optimal case. Let $V_1 \in M(\mathbb{R}L_\infty)$, strictly proper and such that $V_1^* \in M(\mathbb{R}H_\infty)$; so V_1 is anti-stable. Suppose $\| \Gamma_{V_1} \| = \alpha$. Let $\varepsilon > 0$. Then the norm of the Hankel-operator with symbol W_1, defined by

$$W_1 := \frac{1}{\alpha(1+\varepsilon)} V_1 \tag{5.54}$$

is smaller than one:

$$\| \Gamma_{W_1} \| = \| \Gamma_{\frac{1}{\alpha(1+\varepsilon)} V_1} \| = \frac{1}{\alpha(1+\varepsilon)} \| \Gamma_{V_1} \| = \frac{1}{1+\varepsilon} < 1.$$

So we can apply theorem 5.4.1. to W_1. This gives us a parametrization of all matrices $R \in$ $M(\mathbb{R}H_\infty)$ such that

$$\| W_1 - R \| \leq 1.$$

Now suppose R_1 is a matrix that belongs to that set, so $\| W_1 - R_1 \| \leq 1$, with $R_1 \in M(\mathbb{R}H_\infty)$. Then we also have:

$$
\begin{aligned}
\| V_1 - \alpha(1+\varepsilon)R_1 \| &= \| \alpha(1+\varepsilon)W_1 - \alpha(1+\varepsilon)R_1 \| = \\
&= \alpha(1+\varepsilon) \| W_1 - R_1 \| \leq \alpha(1+\varepsilon) = \| \Gamma_{V_1} \| (1+\varepsilon).
\end{aligned}
\tag{5.55}
$$

So $\alpha(1+\varepsilon)R_1$ is a sub–optimal solution to our Nehari–problem. It is therefore possible to construct the set of sub–optimal solutions to the Nehari–problem for the matrix V_1 by multiplying the set of solutions for W_1 by the factor $\alpha(1+\varepsilon)$.

With help of (5.50) we can give a formula for the solution R in theorem 5.4.1. in terms of V_1 and L_1, L_2, L_3 and L_4:

$$
\begin{bmatrix} X_1 \\ X_2 \end{bmatrix} = L \begin{bmatrix} Y \\ I \end{bmatrix} = \begin{bmatrix} L_1 & L_2 \\ L_3 & L_4 \end{bmatrix} \begin{bmatrix} Y \\ I \end{bmatrix} = \begin{bmatrix} L_1 Y + L_2 \\ L_3 Y + L_4 \end{bmatrix}.
$$

So

$$R = V_1 - (L_1 Y + L_2)(L_3 Y + L_4)^{-1}, \qquad Y \in M(\mathbb{R}H_\infty), \| Y \| \leq 1. \tag{5.56}$$

Now we are often not interested in the whole set of solutions, but only in a particular one. Formula (5.56) suggests that the simplest solution to be calculated is given by the choice of $Y = 0$. Then (5.56) becomes

$$R = V_1 - L_2 L_4^{-1}. \tag{5.57}$$

We now summarize the results of this section in the form of an algorithm which calculates a sub–optimal solution to the Nehari–problem, with help of formula (5.57).

Step 0:   Give a matrix $V \in M(\mathbb{R}L_\infty)$, and a tolerance level $\varepsilon > 0$.

Step 1:   Find a realization $[A, B, C, 0]$ of the strictly proper and anti–stable part of V(s):

$$V(s) = V_1(s) + V_2(s)$$

where $V_1(s) = [A,B,C,0]$ with A anti–stable, and $V_2(s)$ a matrix in $M(\mathbb{RH}_\infty)$.

(In our particular case we have:

$$V_1(s) = [-A_c^T, (I+XY)C^T L^{-1/2}, H^{-1/2}B^T, 0]$$
$$V_2(s) = (-H^{-1/2}D^T L^{1/2}) \ ).$$

Step 2:   Calculate $\alpha := \| \Gamma_{V_1} \|$ by the method of section 5.3.

Step 3:   Set $W_1 := \dfrac{1}{\alpha(1+\varepsilon)} V_1$, so $W_1(s) = [A',B',C',0]$, with

$$A' = A, \quad B' = B, \quad C' = \frac{1}{\alpha(1+\varepsilon)} C.$$

Step 4:   Solve the Lyapunov equations

$$A'L_c + L_c A'^T = B'B'^T$$
$$A'^T L_o + L_o A' = C'^T C'.$$

Step 5:   Set $N := (I-L_o L_c)^{-1}$ and define

$$L_2(s) := [A', N^T B', C', 0]$$
$$L_4(s) := [-A'^T, NL_o B', B'^T, I].$$

Step 6:   A sub–optimal solution R_ε to the Nehari–problem is then given by

$$\begin{aligned}
R_\varepsilon(s) &= V_2(s) + \alpha(1+\varepsilon)(W_1(s)-L_2(s)L_4(s)^{-1}) = \\
&= V(s) - \alpha(1+\varepsilon)L_2(s)L_4(s)^{-1}.
\end{aligned} \tag{5.58}$$

Remark that in the algorithm described above two pairs of Lyapunov equations have to be solved, which are highly related. For the computation of $\| \Gamma_{V_1} \|$ we have to solve the Lyapunov equations with the realization $[A,B,C,0]$ of V_1. Let L_c and L_o denote the solutions to these equations:

$$AL_c + L_c A^T = BB^T \tag{5.59}$$

$$A^T L_o + L_o A = C^T C. \tag{5.60}$$

Later on, we have to solve the Lyapunov equations for a realization $[A',B',C',0]$ of W_1. Suppose L'_c and L'_o denote these solutions, then we have

$$A'L'_c + L'_c A'^T = B'B'^T \tag{5.61}$$

$$A'^T L'_o + L'_o A' = C'^T C'. \tag{5.62}$$

Now we know that the realizations $[A,B,C,0]$ of V_1 and $[A',B',C',0]$ of W_1 are connected by the following relations:

$$A' = A; \quad B' = B; \quad C' = \frac{1}{\alpha(1+\varepsilon)} C. \tag{5.63}$$

Substitution of (5.63) in (5.61) and (5.62) gives

$$A L'_c + L'_c A^T = B B^T \tag{5.64}$$

$$A^T L'_o + L'_o A = \left(\frac{1}{\alpha(1+\varepsilon)}\right)^2 C^T C. \tag{5.65}$$

Combination of (5.64) with (5.59), and (5.65) with (5.60) immediately yields

$$L'_c = L_c$$
$$\tag{5.66}$$
$$L'_o = \left(\frac{1}{\alpha(1+\varepsilon)}\right)^2 L_o. \tag{5.67}$$

So we conclude that we only need to solve the Lyapunov equations once, to solve the Nehari–problem in the sub–optimal case. Putting together the algorithm from section 5.3. (for the norm of a Hankel–operator) and the one described above, and using the relations (5.66) and (5.67) between the solutions of the Lyapunov equations, we derive the following algorithm.

Step 0: Give a matrix $V \in M(\mathbb{RL}_\infty)$, and a tolerance level $\varepsilon > 0$.

Step 1: Find a realization of the strictly proper and anti–stable part of $V(s)$:

$$V(s) = V_1(s) + V_2(s),$$

where $V_1(s) = [A,B,C,0]$ with A anti–stable, and $V_2(s)$ a matrix in $M(\mathbb{RH}_\infty)$.

Step 2:   Solve the Lyapunov equations

$$AL_c + L_c A^T = BB^T$$
$$A^T L_o + L_o A = C^T C.$$

Step 3:   Calculate the norm of the Hankel–operator with symbol V:

$$\alpha := \| \Gamma_V \| = \sqrt{\lambda_{max}(L_c L_o)}.$$

Step 4:   Set $N := (I - (\frac{1}{\alpha(1+\epsilon)})^2 L_o L_c)^{-1}$, and define:

$$L_2(s) := [A, N^T B, \frac{1}{\alpha(1+\epsilon)} C, 0] \tag{5.68}$$

$$L_4(s) := [-A^T, (\frac{1}{\alpha(1+\epsilon)})^2 N L_o B, B^T, I]. \tag{5.69}$$

Step 5:   A sub–optimal solution R_ϵ to the Nehari–problem is then given by

$$R_\epsilon(s) = V(s) - \alpha(1+\epsilon) L_2(s) L_4(s)^{-1}. \tag{5.70}$$

This method is used in the MATLAB–function hanneh to solve the Nehari–problem. Given a state–space realization of a matrix $V \in M(\mathbb{RL}_\infty)$, and a tolerance level $\epsilon > 0$, hanneh computes the norm of the Hankel–operator with symbol V, and a sub–optimal solution to the corresponding Nehari–problem, within the desired accuracy. In the computer implementation there is still one improvement; it uses the formula

$$R_\epsilon(s) = V(s) - [A, N^T B, C, 0] * L_4(s)^{-1} \tag{5.71}$$

instead of (5.70). This formula however, is immediately clear from (5.68), (5.70) and the fact that

$$\alpha(1+\epsilon) L_2(s) = \alpha(1+\epsilon) [A, N^T B, \frac{1}{\alpha(1+\epsilon)} C, 0] = [A, N^T B, C, 0].$$

A state–space realization of R_ϵ is then calculated with help of the addition, multiplication and inversion formulae of page ix.

For more information about the MATLAB–function hanneh, we refer to appendix C.

Finally we recall once more the trivial fact that for the solution of the original problem (find an R ∈ M(ℝH$_\infty$) such that ‖ V+R ‖ ≤ ‖ Γ_V ‖ (1+ε)), we have to replace the matrix R we have found with this algorithm by −R.

5.5. An algorithm to compute an optimally robust controller and its stability radius

In the last three sections we have developed all the tools we need for the construction of an algorithm that solves the problem of optimally robust control in the dynamical case. In this section we put all these pieces together to derive such an algorithm.

Suppose P ∈ ℝ$_p^{n*m}$(s), and let [PA,PB,PC,PD] be a minimal realization of P. Assume PA ≠ [ ]. With the method described in theorem 4.1.1. and implemented in the MATLAB−function ncoprfac, we can calculate state−space realizations of an n.r.b.f. (D$_0$,N$_0$) and an n.l.b.f. (Ñ$_0$,D̃$_0$) of P and of matrices Y$_0$,Z$_0$,Ỹ$_0$ and Z̃$_0$ such that (4.7) holds. With the MATLAB−function speccopr we are also able to compute a state−space realization of the matrix V = D$_0^*$Z̃$_0$+N$_0^*$Ỹ$_0$ by the method of theorem 5.2.1.

Given a state−space realization of the matrix V and a tolerance level ε > 0, the MATLAB−function hanneh computes

$$\alpha = \| \Gamma_V \| = \inf_{R \in \mathbb{R}H_\infty^{m*n}(s)} \| V-R \| = \inf_{R \in \mathbb{R}H_\infty^{m*n}(s)} \| V+R \| \tag{5.72}$$

and a state−space realization of a matrix R$_\varepsilon$ ∈ ℝH$_\infty^{m*n}$(s) such that ‖ V−R$_\varepsilon$ ‖ ≤ ‖ Γ_V ‖ (1+ε). The used method is explained in the last section. Now combination of (5.72) and (5.46) immediately yields the following formula for the maximal stability radius w$_g^{-1}$:

$$w_g^{-1} = (1 + \| \Gamma_V \|^2)^{-1/2} = (1 + \alpha^2)^{-1/2}. \tag{5.73}$$

To realize this bound we have to construct a state−space realization of a matrix R ∈ ℝH$_\infty^{m*n}$(s) such that ‖ V+R ‖ = ‖ Γ_V ‖, and substitute this solution to the Nehari−problem in formula (4.46) or (4.56) to derive an optimally robust controller. The MATLAB−function hanneh however, delivers us a state−space realization of a sub−optimal solution R$_\varepsilon$ to the Nehari−problem, i.e. a matrix R$_\varepsilon$ ∈ ℝH$_\infty^{m*n}$(s) such that ‖ V−R$_\varepsilon$ ‖ ≤ ‖ Γ_V ‖ (1+ε). Substituting −R$_\varepsilon$ into (4.56), we find a sub−optimally robust controller C$_\varepsilon$:

$$C_\varepsilon = (Z̃_0-D_0R_\varepsilon)(Ỹ_0+N_0R_\varepsilon)^{-1}. \tag{5.74}$$

At this moment state–space realizations of all the matrices at the right–hand side of (5.74) are available, so it is possible to calculate a state–space realization of C_ε with the addition, multiplication and inversion formulae of page ix.

At this stage we remark that the state–space realizations of the matrices $\tilde{N}_0, \tilde{D}_0, Y_0$ and Z_0 have not been used in the computation of C_ε. Since the matrix V is calculated directly, using the method of section 5.2., these matrices are not explicitly needed any more. With this knowledge the MATLAB–function speccopr is developed. Given a state–space realization of a plant P, it only returns the state–space realizations of the matrices $N_0, D_0, \tilde{Y}_0$ and $\tilde{Z}_0$ with the properties described in theorem 4.1.1., and a state–space realization of V. For more details about the MATLAB–function speccopr we refer to appendix B. We finally remark here that the function speccopr makes the original MATLAB–function ncoprfac superfluous for the algorithm we are now developing.

There is now only one problem left. Because we didn't calculate an exact solution to the Nehari–problem, but a sub–optimal one, we can't expect the controller C_ε to realize the maximal stability radius w_g^{-1} exactly. So to check whether the controller C_ε is indeed sub–optimally robust in the sense that it realizes w_g^{-1} within an accuracy level of ε, we have to calculate the exact stability radius w^{-1} belonging to C_ε. This can be done with formula (4.58). From theorem 4.3.3. we know that for all $R \in \mathbb{R}H_\infty^{m*n}(s)$:

$$w = \left\| \begin{bmatrix} \tilde{Y}_0 - N_0 R \\ \tilde{Z}_0 + D_0 R \end{bmatrix} \right\| = \sqrt{1 + \| V + R \|^2}.$$

So in our case (with $R = -R_\varepsilon$) we have:

$$w = \sqrt{1 + \| V - R_\varepsilon \|^2}. \tag{5.75}$$

Now, $\| V - R_\varepsilon \|^2 \leq \| \Gamma_V \|^2 (1+\varepsilon)^2$, so

$$w = \sqrt{1 + \| V - R_\varepsilon \|^2} \leq \sqrt{1 + \| \Gamma_V \|^2 (1+\varepsilon)^2} \leq$$

$$\leq \sqrt{(1+\varepsilon)^2 + \| \Gamma_V \|^2 (1+\varepsilon)^2} \leq (1+\varepsilon)\sqrt{1 + \| \Gamma_V \|^2} =$$

$$= (1+\varepsilon) \, w_g. \tag{5.76}$$

And because $\frac{1}{1+\varepsilon} > 1-\varepsilon$, this immediately yields

$$w^{-1} \geq \frac{1}{(1+\varepsilon)w_g} = \frac{1}{1+\varepsilon} \, w_g^{-1} > w_g^{-1} (1-\varepsilon). \tag{5.77}$$

So C_ε is indeed a sub–optimal solution to the problem of optimally robust control.

Again we summarize the results of this section in the form of an algorithm.

Step 0:   Give a minimal realization [PA,PB,PC,PD] of a plant $P \in \mathbb{R}_p^{n*m}(s)$, and a tolerance level $\varepsilon > 0$.

Step 1:   Calculate state–space realizations of the matrices $N_0, D_0, \tilde{Y}_0$ and $\tilde{Z}_0$ as defined in theorem 4.1.1. and of V, defined in (5.4).

Step 2:   Compute $\parallel \Gamma_V \parallel$ and a state–space realization of a matrix $R_\varepsilon \in \mathbb{R}H_\infty^{m*n}(s)$ such that

$$\parallel V{-}R_\varepsilon \parallel \le \parallel \Gamma_V \parallel (1{+}\varepsilon)$$

by the method of section 5.4.

Step 3:   Set $w_g^{-1} := (1 + \parallel \Gamma_V \parallel^2)^{-1/2}$.

Step 4:   Compute a state–space realization [CA,CB,CC,CD] of the sub–optimally robust controller C_ε:

$$C_\varepsilon = (\tilde{Z}_0{-}D_0R_\varepsilon)(\tilde{Y}_0{+}N_0R_\varepsilon)^{-1}$$

with the addition, multiplication and inversion formulae of page ix.

The MATLAB–function robstab is the computer implementation of this algorithm. To take the last step, it uses the MATLAB–functions tfadd, tfmult and tfinv, which are implementations of the addition, multiplication and inversion formulae of page ix respectively. More information about these algorithms can be found in appendix E. For the details of the MATLAB–function robstab, we refer to appendix C.

For the sake of completeness we here recall the fact that the Nehari–problem we have to solve in our algorithm doesn't have a unique sub–optimal solution, but a whole set of sub–optimal solutions (see (5.56)). Each one of them gives rise to an other sub–optimally robust controller. For each $\varepsilon > 0$ it is even possible to parametrize the set of all sub–optimal controllers which realize the maximal stability radius w_g^{-1} within a tolerance level of ε. The algorithm we developed however, calculates only one particular compensator.

Although the algorithm of this section delivers a sub–optimal compensator to the problem of optimally robust control, it still solves the problem in a certain sense. We can see this as follows. Given a plant P, we can calculate the maximal achievable stability radius w_g^{-1}; w_g^{-1} only depends upon P. And when a tolerance level $\varepsilon > 0$ is given, we have seen that we can construct a compensator C_ε which realizes a stability radius w^{-1} such that

$$(1-\varepsilon)\ w_g^{-1} \leq w^{-1} \leq w_g^{-1}. \tag{5.78}$$

So the maximal stability radius w_g^{-1} can be approximated to any wished accuracy.

In practice this means the following. Suppose the plant P_λ is a perturbation of P, and $\delta(P,P_\lambda) < w_g^{-1}$. Then we know there exists a compensator C that stabilizes both P and P_λ. But with our algorithm we can also construct such a controller C. Since $\delta(P,P_\lambda) < w_g^{-1}$, there exists an $\varepsilon > 0$ such that $\delta(P,P_\lambda) < w_g^{-1}(1-\varepsilon)$. Choose such an $\varepsilon > 0$. Then the controller C_ε, computed by our algorithm, stabilizes not only P but also P_λ because

$$\delta(P,P_\lambda)\ <\ w_g^{-1}(1-\varepsilon)\ \leq\ w^{-1}.$$

In this way we can say that our algorithm solves the problem of optimally robust control.

Finally we remark that the performance of this algorithm will be discussed in section 6.4. This will then be illustrated by a little design example.

6. REDUCTION OF THE ORDER OF THE COMPENSATOR

In the last chapter we developed an algorithm for the solution of the problem of optimally robust control. Unfortunately the compensator, calculated by this algorithm, is of a very high order. This is the main reason why this solution is hardly applicable in practice. In this chapter we try to overcome this problem. We develop two new algorithms, of which the last one turns out to reduce the order of the optimally robust controller drastically.

In the first section of this chapter we determine the order of the compensator, calculated by our original algorithm. In this way we try to get some insight in the reasons for the inflation of the order of the controller. With this knowledge we derive an algorithm for the computation of an optimally robust controller of a much lower order in the second section. In section 3 we introduce a fully different method which enables us to decrease the order of the solution even much more. Finally, in section 4, we present a little design example to compare the performances of the different algorithms.

6.1. The order of the compensator

Suppose $P \in M(\mathbb{R}_p(s))$ and let [PA,PB,PC,PD] be a minimal realization of the transfermatrix P. Assume $PA \neq [\ ]$. The order of P is then defined as the size of the square–matrix PA, the system–matrix of the plant. Assume that PA is an $n \times n$ matrix. Then we can determine the order of the compensator C, calculated by our algorithm, in terms of n.

First we have to compute a state–space realization of the matrix V, by the method described in theorem 5.2.1. So the system–matrix of V is $-A_c^T$, with $A_c = (PA)-(PB)F$, and F defined as in theorem 5.2.1. Clearly A_c is also an $n \times n$ matrix, so this state–space realization of V has order n.

For this matrix $V \in M(\mathbb{R}L_\infty)$ we then have to solve the Nehari–problem. From formula (5.70) we know that a solution with an accuracy level of ε is given by

$$R_\varepsilon(s) = V(s) - \alpha(1+\varepsilon)L_2(s)L_4(s)^{-1} \tag{6.1}$$

with

$$L_2(s) = [A,N^T B,\frac{1}{\alpha(1+\varepsilon)} C,0]$$

$$L_4(s) = [-A^T,(\frac{1}{\alpha(1+\varepsilon)})^2 NL_0 B,B^T,I]$$

where A is the system–matrix of the anti–stable part of V. Since in our case this matrix is $-A_c^T$, both L_2 and L_4 have order n. From the multiplication formula of page ix, it is immediately clear that the order of the product of two transfermatrices, multiplicated by this method, is the sum of the orders of the single ones. For the addition of two transfermatrices we also have this property: the order of the sum is the sum of the orders. However, when we invert a transfermatrix by the formula of page ix, the order doesn't change. When we use this knowledge to calculate the order of R_ε, the sub–optimal solution to the Nehari–problem (formula (6.1)), we get the following. $L_4(s)$ has order n, and so does $L_4(s)^{-1}$. Since $L_2(s)$ also has order n, the product $\alpha(1+\varepsilon)L_2(s)L_4(s)^{-1}$ has order 2n. So when we add $V(s)$, of order n, to $-\alpha(1+\varepsilon)L_2(s)L_4(s)^{-1}$, the sum $R_\varepsilon(s)$ has order 3n. So we see that the state–space realization of the solution R_ε to the Nehari–problem, calculated by the MATLAB–function hanneh, is of the order 3n.

This solution R_ε to the Nehari–problem is then substituted in formula (5.74) to get a sub–optimally robust controller C_ε:

$$C_\varepsilon = (\tilde{Z}_0 - D_0 R_\varepsilon)(\tilde{Y}_0 + N_0 R_\varepsilon)^{-1} \qquad (6.2)$$

where $D_0, N_0, \tilde{Z}_0$ and $\tilde{Y}_0$ are defined as in theorem 4.1.1. By formula (4.6) we know that A_c is the system–matrix of all these transfermatrices, where $A_c = (PA)-(PB)F$, so all these transfermatrices have order n. Using the order properties of the addition, multiplication and inversion formulae of page ix, described above, we get: the order of $(\tilde{Y}_0 + N_0 R_\varepsilon)^{-1}$ is equal to the order of $(\tilde{Y}_0 + N_0 R_\varepsilon)$, and this is n+(n+3n) = 5n. The order of $(\tilde{Z}_0 - D_0 R_\varepsilon)$ is also n+(n+3n) = 5n. So the order of C_ε, the product of these transfermatrices, is equal to the sum of their orders: 10n. So the order of the state–space realization of the sub–optimally robust controller calculated by the MATLAB–function robstab, is ten times as high as the order of the original system P.

From the exposition above, we can see that there are two reasons for the inflation of the order of the compensator. At first the order is increased when we solve the Nehari–problem. Given a strictly proper $V \in M(\mathbb{R}L_\infty)$, such that $V^* \in M(\mathbb{R}H_\infty)$ (so V is anti–stable) and of order n, our algorithm delivers a solution to the Nehari–problem of order 3n. Secondly the order is increased because we have to substitute this solution in formula (6.2) and use the formulae of page ix, which increase the order very rapidly. We even remark that this is also the main reason why our solution to the Nehari–problem is already of such a high order.

Finally we remark here that the situation would even be worse when we didn't have formula (5.4) or (5.5) for the construction of a state–space realization of the matrix V. With this method we find a state–space realization of the order n. When the realization of the original plant P is minimal, we even know (theorem 5.2.2.) that this realization is also minimal. Therefore an nth order realization of V is the best we can get. When we calculate V by the original formula $V = D_0^* \tilde{Z}_0 - N_0^* \tilde{Y}_0$, where all the transfermatrices on the right–hand side have order n, the order of the state–space realization of V is (n+n)+(n+n) = 4n. And with this realization of V we have to solve the Nehari–problem. The situation is then even much worse. So theorem 5.2.1. already reduces the order of the solution drastically.

6.2. The method of Glover

To decrease the order of our compensator we develop in this section an other method for the solution of the Nehari–problem. In the last section we saw that the method described in paragraph 5.4. was one of the main reasons for the inflation of the order of the compensator. The method of this section is based upon an article of Glover [7], and delivers an exact solution to the Nehari–problem. When the order of the anti–stable matrix $V \in M(\mathbb{RL}_\infty)$, which has to be approximated by a matrix $R \in M(\mathbb{RH}_\infty)$, has order n, the state–space realization of the solution, constructed by this method is of order n–1. In this way it is possible to reduce the order of the state–space realization of the optimally robust controller to 6n–2.

Before we describe the method of Glover, we first have to introduce some new concepts, which will be needed later in this section. We start with the definition of Hankel–singular values.

DEF.6.2.1. Let $V \in M(\mathbb{RL}_\infty)$, and such that $V^* \in M(\mathbb{RH}_\infty)$. Let [A,B,C,D] be a minimal realization of V. Then A is anti–stable. Let L_c and L_o be the unique solutions to the Lyapunov equations

$$AL_c + L_cA^T = BB^T$$
$$A^TL_o + L_oA = C^TC.$$

Suppose the eigenvalues of L_cL_o are arranged in decreasing order, and let $\lambda_i(L_cL_o)$ denote the ith eigenvalue of L_cL_o. Then the *Hankel–singular values* of V(s) are defined as

$$\sigma_i(V(s)) := \sqrt{\lambda_i(L_cL_o)}. \tag{6.3}$$

So by convention $\sigma_i(V(s)) \geq \sigma_{i+1}(V(s))$.

Remark that this definition makes sense because from theorem 5.3.2. and 5.3.3. we know that all the eigenvalues of L_cL_o are real and nonnegative. When the realization [A,B,C,D] of V is minimal one can even prove that all the Hankel–singular values are larger than zero.

With this knowledge we can now introduce the following result about balanced realizations.

THEOREM 6.2.1. Let $V \in M(\mathbb{RL}_\infty)$, and such that $V^* \in M(\mathbb{RH}_\infty)$. Suppose [A,B,C,D] is a *minimal* realization of V, where A is anti–stable. Let $\sigma_1 = \sigma_2 =..= \sigma_r > \sigma_{r+1} \geq \sigma_{r+2} \geq .. \geq \sigma_n > 0$ be the Hankel–singular values of V. Denote

$\Sigma_1 = \text{diag}(\sigma_{r+1}, \sigma_{r+2}, ..., \sigma_n)$ and define the matrix Σ by

$$\Sigma := \begin{bmatrix} \Sigma_1 & 0 \\ 0 & \sigma_1 I_r \end{bmatrix}. \tag{6.4}$$

Then there exists a basis transformation T, such that for the minimal realization of V defined by

$$V = [A', B', C', D'] := [T^{-1}AT, T^{-1}B, CT, D] \tag{6.5}$$

the following statement holds:
The matrix Σ is the solution to both the Lyapunov equations with respect to the realization (A', B', C'), i.e.

$$A'\Sigma + \Sigma A'^T = B'B'^T \tag{6.6}$$

$$A'^T\Sigma + \Sigma A' = C'^TC'. \tag{6.7}$$

A realization $[A', B', C', D']$ with this property is called a *balanced realization* of V.

PROOF The proof is not very difficult and we refer for it to [7, sec.4].

We remark here that the balanced realizations are a quite important tool. With this theorem it is possible to find a realization of the transfermatrix V, for which the controllability and observability gramians have a very simple structure.

We are now ready to state the main result of this chapter. It is an alternative version of a result in [7].

THEOREM 6.2.2. Let $V \in M(\mathbb{RL}_\infty)$, and assume $V^* \in M(\mathbb{RH}_\infty)$. Suppose [A,B,C,D], with A anti–stable, is a minimal and balanced realization of V such that

$$A\Sigma + \Sigma A^T = BB^T$$

$$A^T\Sigma + \Sigma A = C^TC,$$

with $\Sigma = \begin{bmatrix} \Sigma_1 & 0 \\ 0 & \sigma_1 I_r \end{bmatrix}$ as defined in (6.4).

Partition A,B and C conformally with Σ as

$$A = \begin{bmatrix} A_{11} & A_{12} \\ A_{21} & A_{22} \end{bmatrix}, \qquad B = \begin{bmatrix} B_1 \\ B_2 \end{bmatrix}, \qquad C = (C_1, C_2), \tag{6.8}$$

and define

$$\hat{A} := \Gamma^{-1}(\sigma_1^2 A_{11}^T + \Sigma_1 A_{11} \Sigma_1 - \sigma_1 C_1^T U B_1^T) \tag{6.9a}$$

$$\hat{B} := \Gamma^{-1}(\Sigma_1 B_1 - \sigma_1 C_1^T U) \tag{6.9b}$$

$$\hat{C} := C_1 \Sigma_1 - \sigma_1 U B_1^T \tag{6.9c}$$

$$\hat{D} := D - \sigma_1 U \tag{6.9d}$$

where U is a unitary matrix satisfying

$$B_2 = C_2^T U \tag{6.10}$$

and Γ is defined by

$$\Gamma := \Sigma_1^2 - \sigma_1^2 I. \tag{6.11}$$

Then the matrix $R(s) \in M(\mathbb{R}H_\infty)$, defined by

$$R(s) := [\hat{A}, \hat{B}, \hat{C}, \hat{D}] \tag{6.12}$$

is an *exact* solution to the Nehari–problem for the matrix V, i.e. we have

$$\| V - R \| = \| \Gamma_V \|. \tag{6.13}$$

PROOF For the proof of this theorem we refer to [7, sec.6]. The theorem stated above is a combination of the theorems 6.1. and 6.3. in the article of Glover. There is only one difference. In [7] one wants to approximate a system $V \in M(\mathbb{R}H_\infty)$ by an anti–stable one. We want to do it just the other way round: approximate an anti–stable system by a stable one. However, both cases are in essence the same and can be transformed into each other by replacing $V(s)$ by $V(-s)$ and $R(s)$ by $R(-s)$. In our version of theorem 6.2.2. this transformation has already been carried out. □

Combining the theorems 5.2.2., 6.2.1. and 6.2.2. we develop a method to solve the Nehari–problem for the matrix $V(s)$, defined in (4.57) exactly. Given a *minimal* realization of a plant P, we know by the theorems 5.2.1. and 5.2.2. how we can construct a *minimal* realization of the matrix V, belonging to P. The system–matrix of this realization is clearly anti–stable. But then, according to theorem 6.2.1., there exists also a balanced realization of V. And after balancing, application of theorem 6.2.2. yields a construction of a state–space realization of a matrix $R_g(s)$ which is an exact solution to our Nehari–problem:

$$\| \ V - R_g \ \| = \| \ \Gamma_V \ \|.$$

The order of this solution is equal to n–r, the size of the system–matrix $\hat{A}$ of $R_g(s)$, where r is the multiplicity of the largest Hankel–singular value of V. So in general the order will be n–1.

This method for the exact solution of the Nehari–problem has been implemented in the MATLAB–function hannehgl. It makes use of the MATLAB–function ohkapp, which is a standard function in the MATLAB Robust Control Toolbox. This last function solves the problem as stated by Glover (so not our transformed version) conform Glover's method, but without the use of balanced realizations. The method used is given in [14] and bypasses the numerically often ill–conditioned balanced realization step, required in the original method. For the details of the implementation of the function ohkapp we refer to [2].

Now the MATLAB–function hannehgl first transforms our original problem to the one as stated by Glover, then uses the function ohkapp, and finally transforms the output of ohkapp back, to get an exact solution R_g to our original Nehari–problem. The norm of the Hankel–operator with symbol V, which is equal to the largest Hankel–singular value belonging to V, is also computed. In appendix D more information about the MATLAB–function hannehgl can be found.

Analogous to section 5.5. we now develop an algorithm for the computation of an optimally robust controller. This algorithm is completely the same as the one described in paragraph 5.5., apart from one difference: the Nehari–problem is exactly solved with the method of Glover instead of the sub–optimal solution method of section 5.4. So the exact solution R_g to the Nehari–problem can be substituted as $-R_g$ in formula (4.56) to calculate an optimally robust controller:

$$C_g = (\check{Z}_0 - D_0 R_g)(\check{Y}_0 + N_0 R_g)^{-1}. \tag{6.14}$$

As a consequence the maximal stability radius w_g^{-1} is really achieved by this controller. So the method of Glover delivers a true solution to the problem of optimally robust control.

In the resulting computer algorithm, implemented in the MATLAB–function robstgl, there are two differences with the one developed in section 5.5. (the MATLAB–function robstab). First of all no accuracy level $\varepsilon > 0$ is needed, because the Nehari–problem is solved exactly. Secondly the MATLAB–function hanneh is replaced by the MATLAB–function hannehgl. For the details of the function robstgl, we refer to appendix D.

To conclude this section, we calculate the order of the state–space realization of the optimally robust controller, constructed by the MATLAB–function robstgl. Suppose P ϵ $M(\mathbb{R}_p(s))$ and let [PA,PB,PC,PD] be a minimal realization of P. Let PA an n×n matrix, so P has order n. Analogous as in section 6.1. the state–space realization of the matrix V, related to P, and calculated with formula (5.4) is also of order n. For this matrix we solve the

Nehari–problem with Glover's method. In the general case this gives a solution R_g of the order n–1. Substituting R_g into (6.14) we find the following. $(\tilde{Y}_0 + N_0 R_g)^{-1}$ is of the same order as $(\tilde{Y}_0 + N_0 R_g)$ and since $\tilde{Y}_0$ and N_0 are of the order n, this term is of the order n+(n+(n–1)) = 3n–1. Completely analogous we see that the state–space realization of $(\tilde{Z}_0 - D_0 R_g)$ is of the order n+(n+(n–1)) = 3n–1. So the order of the product is (3n–1)+(3n–1) = 6n–2. So with the method of Glover we can construct a state–space realization of an optimally robust controller of the order 6n–2. When we compare this with the result of section 6.1. we draw the conclusion that we have decreased the order of the solution considerably.

6.3. The method of Glover and McFarlane

In this section we describe a fully different method for the computation of a sub–optimally robust controller. It is based upon a recent article [8] of Glover and McFarlane. Given a plant P of order n, they give a method to construct a state–space realization of the order n of a compensator, which turns out to be a sub–optimal solution to our problem of optimally robust control. But before we can derive this result, we need some more theory. Therefore we start with the following lemma. (An alternative version of this lemma is given in [8], but without proof.)

LEMMA 6.3.1. Let $P \in \mathbb{R}_p^{n*m}(s)$, and suppose [A,B,C,D] is a minimal realization of P. Assume $A \neq [\ ]$. Let X, respectively Y, be the unique positive definite solutions to the Algebraic Riccati Equations (4.3) and (4.4). Define as in theorem 4.1.1.

$$A_c := A - BF \quad \text{with } F := H^{-1}(D^T C + B^T X) \qquad (\text{where } H = (I + D^T D)\ )$$
$$A_o := A - KC \quad \text{with } K := (BD^T + YC^T)L^{-1} \qquad (\text{where } L = (I + DD^T)\ ).$$

Then the matrices $(I+YX)^{-1}Y$ and $(I+XY)^{-1}X$ satisfy the following Lyapunov equations:

$$(I+YX)^{-1}YA_c^T + A_c(I+YX)^{-1}Y = -BH^{-1}B^T \qquad (6.15)$$
$$(I+XY)^{-1}XA_o + A_o^T(I+XY)^{-1}X = -C^T L^{-1}C. \qquad (6.16)$$

PROOF i) First we prove (6.15). Using the fact that $L^{-1}D = DH^{-1}$ (formula (5.12)) we derive from the definitions of A_c and A_o:

$$A_c^T - A_o^T = (A-BF)^T - (A-KC)^T = C^T K^T - F^T B^T =$$

$$= C^T L^{-1}(DB^T + CY) - (C^T D + XB)H^{-1}B^T =$$
$$= C^T L^{-1} DB^T + C^T L^{-1} CY - C^T DH^{-1}B^T - XBH^{-1}B^T =$$
$$= C^T L^{-1} CY - XBH^{-1}B^T. \tag{6.17}$$

Pre–multiplication of (6.17) with Y, addition of the term $-BH^{-1}B^T$ on both right– and left–hand side, and a little rearranging yields

$$YA_c^T - YA_o^T - BH^{-1}B^T - YC^T L^{-1}CY \;=\; -BH^{-1}B^T - YXBH^{-1}B^T =$$
$$= -(I+YX)BH^{-1}B^T. \tag{6.18}$$

Now from (4.13) we know that

$$A_o Y = -YA_o^T - YC^T L^{-1}CY - BH^{-1}B^T. \tag{6.19}$$

Substituting (6.19) in (6.18), we get

$$YA_c^T + A_o Y = -(I+YX)BH^{-1}B^T, \tag{6.20}$$

and pre–multiplication with $(I+YX)^{-1}$ yields:

$$(I+YX)^{-1}YA_c^T + (I+YX)^{-1}A_o Y = -BH^{-1}B^T. \tag{6.21}$$

From (5.20) we know that $(I+XY)A_o^T = A_c^T(I+XY)$, so $A_o(I+YX) = (I+YX)A_c$ and $(I+YX)^{-1}A_o = A_c(I+YX)^{-1}$. Substitution of this equality in (6.21) gives

$$(I+YX)^{-1}YA_c^T + A_c(I+YX)^{-1}Y = -BH^{-1}B^T,$$

and this exactly formula (6.15).

ii) The proof of (6.16) is almost completely analogous. When we transpose formula (6.17), change the roles of X and Y, and of $BH^{-1}B^T$ and $C^T L^{-1}C$, and use (4.8) instead of (4.13), the proof of (6.16) follows immediately. The details are therefore omitted. □

With help of this lemma it is possible to reduce the algorithm of section 5.3. for the computation of the maximal stability radius w_g^{-1} to a very simple formula. We already know that the maximal stability radius depends only upon the plant P. The next theorem describes how.

THEOREM 6.3.2. Let $P \in \mathbb{R}_p^{n*m}(s)$, and suppose $[A,B,C,D]$ is a minimal realization of P. Assume $A \neq [\ ]$. Let X, respectively Y, be the unique positive definite solutions to the Algebraic Riccati Equations (4.3) and (4.4).

Then the maximal stability radius w_g^{-1}, belonging to P, i.e. the radius of the largest ball around P that can be stabilized by one compensator C, is given by:

$$w_g^{-1} = \frac{1}{\sqrt{1 + \lambda_{max}(XY)}}. \tag{6.22}$$

PROOF First we remark that in this proof we feel free to use the different matrix definitions given in the theorems 4.1.1. and 5.2.1. Now to prove (6.22) we simply carry out the algorithm for the computation of the maximal stability radius w_g^{-1} of section 5.3.

Let $P = [A,B,C,D] \in \mathbb{R}_p^{n*m}(s)$. Then first we have to find a state–space realization of the strictly proper anti–stable part $V_1(s)$ of the matrix $V(s)$, associated with P, as defined in (4.57). According to theorem 5.2.1. and because A_c is a stable matrix, this is given by

$$V_1(s) = [-A_c^T,(I+XY)C^TL^{-1/2},H^{-1/2}B^T,0].$$

With this realization, we have to solve the Lyapunov equations (5.42) and (5.43). This yields:

$$-A_c^T L_c - L_c A_c = (I+XY)C^TL^{-1}C(I+YX) \tag{6.23}$$

$$-A_c L_o - L_o A_c^T = BH^{-1}B^T. \tag{6.24}$$

When we compare (6.24) with (6.15), we immediately see that the solution L_o to (6.24) is equal to

$$L_o = (I+YX)^{-1}Y. \tag{6.25}$$

To solve (6.23) we have to do some more work. First we pre–multiply (6.23) with $-(I+XY)^{-1}$ and post–multiply it with $(I+YX)^{-1}$. This gives

$$(I+XY)^{-1}A_c^T L_c(I+YX)^{-1} + (I+XY)^{-1}L_c A_c(I+YX)^{-1} = -C^TL^{-1}C. \tag{6.26}$$

We now recall formula (5.20): $(I+XY)A_o^T = A_c^T(I+XY)$. From this it follows immediately that $(I+XY)^{-1}A_c^T = A_o^T(I+XY)^{-1}$, and after transposition that $A_c(I+YX)^{-1} = (I+YX)^{-1}A_o$. Substitution of these formulae in (6.26) yields

$$A_o^T(I+XY)^{-1}L_c(I+YX)^{-1} + (I+XY)^{-1}L_c(I+YX)^{-1}A_o = -C^TL^{-1}C. \tag{6.27}$$

Combining (6.27) with (6.16) we get

$$(I+XY)^{-1}L_c(I+YX)^{-1} = (I+XY)^{-1}X. \tag{6.28}$$

Pre–multiplication of (6.28) with (I+XY) and post–multiplication with (I+YX) gives

$$L_c = X(I+YX). \tag{6.29}$$

With help of (6.25) and (6.29) we can now calculate L_cL_o:

$$L_cL_o = X(I+YX)(I+YX)^{-1}Y = XY. \tag{6.30}$$

So we get $\| \Gamma_V \| = \sqrt{\lambda_{max}(L_cL_o)} = \sqrt{\lambda_{max}(XY)}$, and finally with formula (5.46):

$$w_g^{-1} = (1 + \| \Gamma_V \|^2)^{-1/2} = (1 + \lambda_{max}(XY))^{-1/2},$$

which is exactly formula (6.22).

This completes the proof. □

Before we introduce the result of Glover and McFarlane we recall some definitions from chapter 4. In formula (4.45) we defined for $\varepsilon > 0$ a neighborhood of a plant $P_0 \in \mathbb{R}^{n*m}(s)$ in the gap–topology:

$$K(P_0,\varepsilon) = \{ P \in \mathbb{R}^{n*m}(s) \mid \delta(P_0,P) < \varepsilon \}. \tag{6.31}$$

However, it was also possible to describe a neighborhood of P_0 with help of perturbations of its n.r.b.f. (D_0,N_0). According to (4.47) we defined for $\varepsilon > 0$:

$$R(P_0,\varepsilon) = \{ P = (N_0+\Delta_n)(D_0+\Delta_d)^{-1} \in \mathbb{R}^{n*m}(s) \mid \| \begin{bmatrix} \Delta_d \\ \Delta_n \end{bmatrix} \| < \varepsilon \}, \tag{6.32}$$

where (D_0,N_0) is an n.r.b.f. of P_0. Glover and McFarlane used this last description to tackle the problem of robust stabilization. They derived the following result.

THEOREM 6.3.3. Let $P \in \mathbb{R}_p^{n*m}(s)$ and suppose [A,B,C,D] is a minimal realization of P. Assume $A \neq [\ ]$. Let X, respectively Y, be the unique positive definite solutions to the Algebraic Riccati Equations (4.3) and (4.4). Then the following statements hold:

i) The largest radius r, such that R(P,r) can be stabilized by one compensator C is given by

$$r_{max} = (1 + \lambda_{max}(XY))^{-1/2}. \qquad (6.33)$$

ii) Let $\varepsilon > 0$ and define $\gamma := \dfrac{1}{r_{max}} (1+\varepsilon)$. Define as before

$$A_c := A-BF \quad \text{with} \quad F := H^{-1}(D^TC+B^TX) \qquad (\text{where } H = (I+D^TD)\ )$$

and

$$W_1 := I + (XY-\gamma^2 I). \qquad (6.34)$$

Then the controller C, given by

$$C = [A_c+\gamma^2(W_1^{-1})^TYC^T(C+DF),\gamma^2(W_1^{-1})^TYC^T,B^TX,-D^T], \qquad (6.35)$$

stabilizes $R(P,r_{max} \dfrac{1}{1+\varepsilon})$. This controller C is of the *same order* as the original plant P.

PROOF For the proof we refer to [8, pp.825–826].

Now, when we compare formula (6.33) for r_{max} with (6.22) for w_g^{-1}, we see that they are equal:

$$r_{max} = w_g^{-1}. \qquad (6.36)$$

So for both the characterizations of a neighborhood of P, the maximal stability radius is the same. This result is not very surprising when we recall the result of theorem 4.2.3. There we proved that for $0 < \varepsilon \leq 1$, $K(P,\varepsilon) = R(P,\varepsilon)$, so these neighborhoods are the same. But then the controller that stabilizes $R(P,r_{max} \dfrac{1}{1+\varepsilon})$ should also stabilize $K(P,w_g^{-1} \dfrac{1}{1+\varepsilon})$.

The validity of this last statement we can see as follows. Since C stabilizes P, $C \in S(P)$. Now let $(\tilde{D}_0,\tilde{N}_0)$ be an n.r.b.f. of P and (Y_0,Z_0) matrices such that (4.38) holds. Then we know that there exists an $R \in M(\mathbb{R}H_\infty)$ such that

$$C = (Y_0-R\tilde{N}_0)^{-1}(Z_0+R\tilde{D}_0) \qquad (6.37)$$

(see formula (4.39)). Now C stabilizes $R(P,w_g^{-1}\frac{1}{1+\epsilon})$, so according to theorem 4.2.1. we have

$$\| (Y_0-R\tilde{N}_0),(Z_0+R\tilde{D}_0) \| \leq w_g(1+\epsilon). \tag{6.38}$$

But then formula (4.42) gives that the stability radius w^{-1} of C, in terms of the gap–topology is

$$w^{-1} = \| (Y_0-R\tilde{N}_0),(Z_0+R\tilde{D}_0) \|^{-1} \geq w_g^{-1}\frac{1}{1+\epsilon} > w_g^{-1}(1-\epsilon). \tag{6.39}$$

So C also stabilizes $K(P,w_g^{-1}(1-\epsilon))$ and we see that the controller C, defined by (6.35) is a sub–optimal solution to the problem of optimally robust control. The maximal stability radius w_g^{-1} can be achieved by this controller to any wished accuracy by choosing ϵ appropriately small. The great advantage of this controller, however, is its order: the order of the state–space realization of C given in (6.35) is equal to the order of the original plant P. With this conclusion the problem of order–reduction seems to be solved.

This method of Glover and McFarlane for the solution of the problem of optimally robust control has been implemented for computer purposes in the MATLAB–function rbstglfar. Given a minimal realization of a plant $P \in \mathbb{R}_p^{n*m}(s)$ and a tolerance level $\epsilon > 0$, rbstglfar calculates the maximal stability radius w_g^{-1} belonging to P, and a state–space realization of a controller C that achieves this bound within the desired accuracy level, with help of the formulae (6.22) and (6.35). The order of this compensator is equal to the order of the given plant.

Remark that for this computation only two Riccati–equations have to be solved. In contrast with the other methods no Lyapunov equations have to be solved (we solved them analytically in lemma 6.3.1.). Explicit state–space realizations of Bezout factorizations are also not needed any more. In this way the MATLAB–function rbstglfar is a very simple algorithm for the solution of the problem of optimally robust control. For the details of this implementation, we refer to appendix D.

To conclude this section we finally remark that the performance of the MATLAB–function rbstglfar will be one of the subjects of the next section.

6.4. The performance of the algorithms

In chapter 5 and the last two sections of this chapter, we developed three different algorithms for the solution of the problem of optimally robust control. In this section we will compare them to find out what the advantages and disadvantages of these methods are. This will be done with help of a little design example.

We choose a plant $P \in M(\mathbb{R}_p(s))$ with transferfunction $P(s) = \frac{1}{s-1}$. Then P is an unstable plant (it has a pole in +1), and a state–space realization of P is given by P(s) = [A,B,C,D] with A = B = C = 1 and D = 0. For this plant we computed the maximal stability radius w_g^{-1} and a state–space realization of a (sub)–optimally robust controller C in five different ways:

i)   with the MATLAB–function robstab; tolerance level $\varepsilon = 10^{-5}$;
ii)   with the MATLAB–function robstab; tolerance level $\varepsilon = 10^{-8}$;
iii) with the MATLAB–function robstgl;
iv)   with the MATLAB–function rbstglfar; tolerance level $\varepsilon = 10^{-5}$;
v)   with the MATLAB–function rbstglfar; tolerance level $\varepsilon = 10^{-8}$.

The results of these computations are listed in appendix F.

Comparing these results, we immediately see that all the methods give the same value for the maximal achievable stability radius w_g^{-1}; $w_g^{-1} = 0.3827$. This is exactly what we expected. The three methods calculate w_g^{-1} in almost the same way, they differ only in the way they compute a (sub)–optimally robust controller that realizes this bound. The example also illustrates the order inflation of the different methods. Our original plant P is of order 1. The compensators calculated by the MATLAB–function robstab are of the order 10, by robstgl of the order 4 (4=6–2), and by rbstglfar of order 1, in accordance with our predictions in the last sections.

In the solutions for the sub–optimal case (calculated by robstab and rbstglfar), we also encounter an other phenomenon: the controller turns out to be highly dependent upon the tolerance level ε. When a tolerance level ε has to be achieved, the order of magnitude of the compensators is about $1/\varepsilon$. So, when the maximal stability radius has to be approximated very accurately, the compensators calculated this way have very large poles and zeros. This is very unsatisfactory because such controllers are hardly applicable in practice.

The reason for this behavior is quite clear. Take for example the function robstab. In this algorithm we have to solve a Nehari–problem before we can compute a sub–optimally robust controller. For the computation of a sub–optimal solution to the Nehari–problem, we have to calculate (see the second algorithm of section 5.4., step 4):

$$N = (I - (\frac{1}{\alpha(1+\varepsilon)})^2 \, L_o L_c)^{-1}, \tag{6.40}$$

where $\alpha = \sqrt{\lambda_{max}(L_c L_o)} > 0$. But then α^2 is also an eigenvalue of $L_o L_c$. Suppose x is an eigenvector of $L_c L_o$ corresponding to eigenvalue α^2. Then $L_c L_o x = \alpha^2 x$, and $L_o L_c L_o x = \alpha^2 L_o x$. So $L_o x$ is an eigenvector of $L_o L_c$ corresponding to the eigenvalue α^2, provided that $L_o x \neq 0$. But this is quite clear from the fact that $L_c L_o x = \alpha^2 x \neq 0$. So α^2 is indeed an eigenvalue of $L_o L_c$.

Now define $M := (I-(\frac{1}{\alpha(1+\epsilon)})^2 L_oL_c)$. Then $N = M^{-1}$. From the argument above it is clear that $(\frac{1}{1+\epsilon})^2$ is an eigenvalue of $(\frac{1}{\alpha(1+\epsilon)})^2 L_oL_c$, so $1-(\frac{1}{1+\epsilon})^2$ is an eigenvalue of M. So, when ϵ tends to zero, the matrix M becomes almost singular. For small tolerance levels $\epsilon > 0$, the matrix N is the inverse of an almost singular matrix. The computation of such an inverse is numerically unstable, and it is obvious that in this way the inverse of the tolerance level, ϵ^{-1}, gets into our solution. So the algorithm robstab is numerically not very satisfactory.

When we look at the function rbstglfar, we see that almost the same thing happens. Here we have to compute the inverse of the matrix

$$W_1 = I + (XY-\gamma^2I), \tag{6.41}$$

where $\gamma^2 = (1+\lambda_{max}(XY))(1+\epsilon)^2$ (see formulae (6.33) and (6.34)). Let $x \neq 0$ be an eigenvector of XY corresponding to the largest eigenvalue λ_{max} of XY. Then we have:

$$\begin{aligned} W_1x &= (I + (XY-\gamma^2I))x = x + \lambda_{max}x - \gamma^2x = \\ &= ((1+\lambda_{max}) - (1+\lambda_{max})(1+\epsilon)^2)x = \\ &= (1+\lambda_{max})(-2\epsilon-\epsilon^2)x. \end{aligned} \tag{6.42}$$

So $(1+\lambda_{max})(-2\epsilon-\epsilon^2)$ is an eigenvalue of W_1, and when ϵ tends to zero, this eigenvalue does too. For small tolerance levels $\epsilon > 0$, the matrix W_1 becomes almost singular, and because we have to compute the inverse of W_1 to determine a sub-optimally robust controller C, we get into the same numerical problems as before. From the exposition above it is quite obvious that in this case the solution to the problem of optimally robust control depends upon the tolerance level ϵ, in the way we have seen it in our little design example.

Fortunately, our third algorithm, the function robstgl, overcomes these numerical difficulties. In this algorithm, the method of Glover is used to calculate an exact solution to the Nehari-problem and no tolerance level is used. So in this case we can't encounter the numerical problems of the other two algorithms, because they were caused by this tolerance level. When we look at the results delivered by this algorithm for our example, we see that the order of magnitude of the compensator is equal to the order of magnitude of the original plant. This result is very satisfactory. When we study the method of Glover more carefully we come to the conclusion that also other numerical difficulties are not very likely to happen. The only problem is the matrix Γ, defined in (6.11) by

$$\Gamma = \Sigma_1^2 - \sigma_1^2I,$$

which has to be inverted. This matrix can become almost singular when the largest Hankel-singular value is almost equal to the largest but one. This is not very likely to happen, and fully dependent upon the transfermatrix of the original plant P. We could say that in this case the problem is ill-conditioned.

Because of the numerical difficulties mentioned before, we can ask ourselves whether the compensators calculated by the MATLAB–functions robstab and rbstglfar still achieve the maximal stability radius w_g^{-1} within the desired accuracy. Because of rounding errors, the sub–optimality could be endangered. To check whether the calculated solutions are still (sub)–optimally robust, we simply compute the stability radius w^{-1} belonging to the plant–compensator pair (P,C) by the method given in theorem 3.2.5. First we calculate an n.r.b.f. (D_0,N_0) of P and an n.l.b.f. $(\tilde{N}_0,\tilde{D}_0)$ of C. Since these factorizations are normalized, we have

$$\left\| \begin{bmatrix} D_0 \\ N_0 \end{bmatrix} \right\| = \| (\tilde{D}_0,\tilde{N}_0) \| = 1,$$

and the formula for the stability radius w^{-1} as given in theorem 3.2.5. becomes

$$w^{-1} = \| (\tilde{D}_0 D_0 + \tilde{N}_0 N_0)^{-1} \|^{-1}. \tag{6.43}$$

A state–space realization of the matrix $U := \tilde{D}_0 D_0 + \tilde{N}_0 N_0$ is easily obtained from the different algorithms we already developed. Normalized right– and left–Bezout factorizations can be constructed with help of the MATLAB–function ncoprfac. A state–space realization of the matrix U^{-1} is then computed with the functions tfmult, tfadd, and tfinv, which are based upon the formulae of page ix.

When we carry out this procedure for the compensator calculated with the MATLAB–function robstab with tolerance level $\varepsilon = 10^{-5}$, we get the following. U^{-1} is in this case a transferfunction (P is a SISO–system) and has a number of poles in the RHP. Theoretically these poles have to be canceled out by zeros, and we see that this happens in our example. When we also carry out a rough pole–zero cancellation in the LHP, and then calculate the norm of U^{-1} with help of the MATLAB–function hinfn (see [1]), we find $\| U^{-1} \| = 2.6132$, and $w^{-1} = 1/\| U^{-1} \| = 0.3827$. This is exactly the value of w_g^{-1} we already calculated. So the compensator we have calculated is still sub–optimally robust.

Using the same procedure for the compensator computed by the MATLAB–function rbstglfar (with tolerance level $\varepsilon = 10^{-8}$) we observe almost the same behavior as before. The transfermatrix U^{-1} has, after a rough pole–zero cancellation , only very large poles and zeros. Its norm turns out to be equal to 2.6131. So again we find $w^{-1} = 0.3827$.

However, when we check the optimality of the solution calculated by the function robstgl, we find immediately a transferfunction U^{-1} with all its poles in the LHP. So it is possible to calculate the H_∞–norm without any pole–zero cancellation. The value of w^{-1} turns out to be equal to 0.3827, as expected. Because in this case no pole–zero cancellation is necessary, we conclude again that the MATLAB–function robstgl is numerically the most reliable one.

To conclude this section we finally summarize the results we have found. First of all the MATLAB–function robstab turns out to be not very useful in practice. It gives solutions of a very high order and the order of magnitude of the compensator depends upon the given tolerance level. The function robstgl, on the other hand, is numerically very reliable, but the order of the compensator calculated by this algorithm is still rather high. This problem can be overcome by the function rbstglfar, which calculates a controller of the same order as the plant. This function however, has the disadvantage that it has the same undesirable numerical properties as the function robstab and produces solutions with very large poles and zeros. So we can give the following rule of the thumb for the use of the different algorithms. When numerical reliability is more important than the order of the compensator, use the MATLAB–function robstgl. When a lower order is preferred, use the function rbstglfar. The disadvantage of a compensator with very large poles and zeros has then to be taken into account. So the choice of the method to use is a trade–off between undesirable numerical properties and the order of the solution.

7. CONCLUSIONS

In this final chapter we draw our conclusions. In the first section we will summarize the results we have found so far. Later on, in the second paragraph, we give a short outline for eventual future developments in this field.

7.1. Summary

In this paper we achieved the following:

— We described the problem of robust stabilization, and introduced the gap–topology to measure distances between plants. We saw that the gap–topology is compatible with the problem of robust stabilization.
— We gave sufficient conditions for robust BIBO–stabilization in the gap–metric.
— With help of the guaranteed bounds for robust stabilization we were able to find a solution to the problem of optimally robust control in the gap–topology. By maximizing the bound for robust stabilization, we found a (theoretical) method to calculate, for each plant P $\in$ M($\mathbb{R}$(s)), the maximal stability radius belonging to it (i.e. the radius of the largest ball around P that can be stabilized by one controller C), and a controller C that realizes this bound.

Most of the results, stated above, are due to Zhu (see [19]). In the derivation of them, right– and left–Bezout factorizations turned out to play a central role. These factorizations appeared to be a crucial tool in the development of the theory.

In the second part of this paper, we made these results more concrete, and developed three algorithms for the computation of an optimally robust controller and the maximal stability radius for the case P $\in$ M($\mathbb{R}_p$(s)). (In this case, P has a state–space realization P = [PA,PB,PC,PD].) Here we derived the following results:

— Given a state–space realization of a plant P $\in$ M($\mathbb{R}_p$(s)), we gave state–space realizations of a normalized doubly–Bezout factorization of P, i.e. we constructed state–space realizations of a normalized right– and left–Bezout factorization of P ((D_0,N_0) and ($\tilde{N}_0,\tilde{D}_0$) respectively) and of matrices $Y_0,Z_0,\tilde{Y}_0$ and $\tilde{Z}_0$ such that

$$\begin{bmatrix} -Z_0 & Y_0 \\ \tilde{D}_0 & \tilde{N}_0 \end{bmatrix} \begin{bmatrix} -N_0 & \tilde{Y}_0 \\ D_0 & \tilde{Z}_0 \end{bmatrix} = \begin{bmatrix} I & 0 \\ 0 & I \end{bmatrix}. \tag{7.1}$$

— We showed that the problem of optimally robust control can be reduced to a

Nehari–problem for a matrix V, related to the original plant P. Given a minimal state–space realization of P, we gave a method to construct a minimal state–space realization of this matrix V.

— With help of this matrix V, associated with P, we developed an algorithm that computes the maximal stability radius belonging to P, and a state–space realization of a controller C that realizes this bound. The controller C was calculated in two different ways:

i)   with help of a sub–optimal solution to the Nehari–problem, found with the method of Francis (see chapter 5). This finally led to a sub–optimally robust compensator of an order ten times as high as the order of the original plant P.

ii) by the method of Glover, which uses an exact solution to the related Nehari–problem. This yields an optimally robust controller of an order almost six times as high as the one of the original plant.

— Finally we gave a direct method to compute the maximal stability radius, belonging to a plant P, given a minimal state–space realization of it. We showed that the sub–optimally robust controller, based upon an other metric for plants, and given by Glover and McFarlane in [8], is also sub–optimally robust in our setting, i.e. in the gap–topology. The order of this compensator is equal to the order of the plant.

— We concluded our exposition with a comparison of the three algorithms. We saw that the method of Glover was numerically the most reliable one, but with the disadvantage of a high order compensator. The method of Glover and McFarlane did overcome this problem at the cost of undesirable numerical effects.

7.2. An outline for future developments

In section 6.4., where we compared the performances of the algorithms, we saw that they all have their own advantages and disadvantages. But unfortunately there was no algorithm that combined all the advantages, yielding a compensator of the same (or lower) order as the plant in a numerically stable way, and without any drawbacks. Of course now the question arises whether such a compensator exists, and if so, how to calculate it. For a very special case, the first question can be answered affirmatively.

THEOREM 7.2.1. Let $P \in \mathbb{R}_p(s)$ a first order, strictly proper system with transferfunction

$$P(s) = \frac{\beta}{s - \alpha}, \qquad \text{with } \beta \neq 0. \qquad (7.2)$$

Then the maximal stability radius belonging to P is

$$w_g^{-1} = \frac{|\beta|}{\sqrt{2(\alpha^2 + \beta^2 + \alpha\sqrt{\alpha^2+\beta^2})}} . \qquad (7.3)$$

And an optimally robust controller of the order *zero*, that realizes this bound, is given by:

$$K = \frac{\alpha + \sqrt{\alpha^2 + \beta^2}}{\beta} . \tag{7.4}$$

PROOF i) First we prove formula (7.3) for the maximal stability radius w_g^{-1}. To do this, we simply apply theorem 6.3.2.

A minimal state–space realization of P is given by P = [A,B,C,D] with A = α, B = β, C = 1 and D = 0. First we have to solve two Riccati equations (formulae (4.3) and (4.4)), which in this case are scalar:

$$\alpha x + x\alpha - \beta^2 x^2 + 1 = 0 \tag{7.5}$$
$$\alpha y + y\alpha - y^2 + \beta^2 = 0. \tag{7.6}$$

The positive (definite) solutions to (7.5) and (7.6) are given by

$$x = \frac{\alpha + \sqrt{\alpha^2 + \beta^2}}{\beta^2} ; \qquad\qquad y = \alpha + \sqrt{\alpha^2 + \beta^2}. \tag{7.7}$$

Now we know by formula (6.22) that w_g^2 is given by

$$w_g^2 = 1 + \lambda_{max}(xy) = 1 + xy = 1 + \frac{(\alpha + \sqrt{\alpha^2+\beta^2})^2}{\beta^2} =$$
$$= 2 \left\{ \frac{\alpha^2 + \beta^2 + \alpha\sqrt{\alpha^2+\beta^2}}{\beta^2} \right\}.$$

So, for the maximal stability radius w_g^{-1} we have

$$w_g^{-1} = \frac{|\beta|}{\sqrt{2(\alpha^2 + \beta^2 + \alpha\sqrt{\alpha^2+\beta^2})}} ,$$

and this is exactly formula (7.3).

ii) To prove that the controller K, defined in formula (7.4) is indeed optimally robust, we show that the stability radius w^{-1} belonging to the plant–compensator pair (P,K) is equal to w_g^{-1}. To do so, we compute this radius w^{-1}, by the method given in theorem 3.2.5.

First we construct an n.r.b.f. (D_0, N_0) of P. Since we already solved the Riccati equations, we only have to substitute the solutions of them in the state–space formulae of theorem 4.1.1. This yields

$$f = \beta x = \frac{\alpha + \sqrt{\alpha^2 + \beta^2}}{\beta}; \qquad\qquad a_c = \alpha - \beta f = - \sqrt{\alpha^2 + \beta^2}.$$

(The capitals are replaced by lower case letters, because they all represent scalars.) And using formula (4.6), we get:

$$D_0(s) = 1 - \frac{\beta f}{s - a_c} = 1 - \frac{\alpha + \sqrt{\alpha^2 + \beta^2}}{s + \sqrt{\alpha^2 + \beta^2}} = \frac{s - \alpha}{s + \sqrt{\alpha^2 + \beta^2}} . \qquad\qquad (7.8)$$

$$N_0(s) = \frac{\beta}{s - a_c} = \frac{\beta}{s + \sqrt{\alpha^2 + \beta^2}} . \qquad\qquad (7.9)$$

With help of theorem 4.1.2. it is possible to construct an n.l.b.f. $(\tilde{N}_0, \tilde{D}_0)$ of the constant transfermatrix K. Application of formula (4.34) yields:

$$\tilde{D}_0 = (I + KK^T)^{-1/2} = (1 + (\frac{\alpha + \sqrt{\alpha^2 + \beta^2}}{\beta})^2)^{-1/2} =$$

$$= \frac{| \beta |}{\sqrt{2(\alpha^2 + \beta^2 + \alpha \sqrt{\alpha^2 + \beta^2})}} . \qquad\qquad (7.10)$$

$$\tilde{N}_0 = (I + KK^T)^{-1/2}K =$$

$$= \frac{\alpha + \sqrt{\alpha^2 + \beta^2}}{\beta} \frac{| \beta |}{\sqrt{2(\alpha^2 + \beta^2 + \alpha \sqrt{\alpha^2 + \beta^2})}} . \qquad\qquad (7.11)$$

We now define, in conformity with (3.7), $A_0 := \begin{bmatrix} D_0 \\ N_0 \end{bmatrix}$, $B_0 := (\tilde{D}_0, \tilde{N}_0)$ and $U_0 := B_0 A_0$. Because the Bezout factorizations are both normalized, we have

$$\| A_0 \| = \| \begin{bmatrix} D_0 \\ N_0 \end{bmatrix} \| = 1,$$

$$\| B_0 \| = \| (\tilde{D}_0, \tilde{N}_0) \| = 1.$$

So the formula for w^{-1}, the stability radius of the plant–compensator pair (P,K) becomes (see theorem 3.2.5.)

$$w^{-1} = (\| A_0 \| \| B_0 \| \| U_0^{-1} \|)^{-1} = \| U_0^{-1} \|^{-1}. \qquad\qquad (7.12)$$

Now we compute U_0:

$$U_0 = \tilde{D}_0 D_0 + \tilde{N}_0 N_0 =$$

$$= \frac{|\beta|}{\sqrt{2(\alpha^2 + \beta^2 + \alpha\sqrt{\alpha^2+\beta^2})}} \left\{ \frac{s - \alpha}{s + \sqrt{\alpha^2+\beta^2}} + \frac{\alpha + \sqrt{\alpha^2+\beta^2}}{s + \sqrt{\alpha^2+\beta^2}} \right\} =$$

$$= \frac{|\beta|}{\sqrt{2(\alpha^2 + \beta^2 + \alpha\sqrt{\alpha^2+\beta^2})}} \left[\frac{s + \sqrt{\alpha^2+\beta^2}}{s + \sqrt{\alpha^2+\beta^2}} \right] =$$

$$= \frac{|\beta|}{\sqrt{2(\alpha^2 + \beta^2 + \alpha\sqrt{\alpha^2+\beta^2})}} = w_g^{-1}, \tag{7.13}$$

where the last equality obviously follows from formula (7.3) for w_g^{-1}. With (7.13) we see immediately that $U_0^{-1} = w_g$, and so

$$w^{-1} = \| U_0^{-1} \|^{-1} = w_g^{-1}.$$

So the stability radius w^{-1} belonging to the plant–compensator pair (P,K) is equal to the maximal achievable stability radius w_g^{-1}, and by consequence, K is an optimally robust controller for P.

This completes the proof. □

For the very special case of theorem 7.2.1. it is possible to calculate an optimally robust controller of a *lower* order than the original plant, in a very simple and numerically stable way. Remark however that the proof of the optimality of K was not constructive. In a certain sense, we introduced an apparently arbitrary compensator K, which turned out to be optimally robust. Of course, this is not the plain truth. We constructed K in such a way that the matrix U_0 (a scalar in this case) was constant, and this K turned out to be optimally robust.

Naturally, now the question arises: can this result be generalized to higher order systems? The answer is ambiguous. When n is the order of the original plant, Glover and McFarlane have claimed that in theory there must exist an optimally robust controller of an order not greater than n–1 (see [8, p.827]). The argument for this claim is the following. For the H_∞–control problem, which is highly related to our problem, almost the same question arose a few years ago. This question, however, has been solved in the mean time, and it turned out that we can construct a controller of a lower order than of the original plant in this case. So we expect this also to be possible for the robust stabilization problem.

But if such a compensator really exists, how should it be calculated in a numerically stable way? And how can we construct a state–space realization of such a controller? These questions have not been answered yet, and further research in this area seems to be worthwile. In this way it should be possible to develop a numerically stable algorithm that computes an optimally robust controller of a lower order than of the original plant. The results of such an algorithm are then also very good applicable in practice.

APPENDIX A

This appendix contains function–descriptions of the MATLAB–functions gap and dirgap.

gap

<u>Purpose</u> :
Computation of the gap between two systems.

<u>Synopsis</u> :
[g1,g2,dist] = gap(A1,B1,C1,D1,A2,B2,C2,D2)

<u>Description</u> :
Given minimal realizations [A1,B1,C1,D1] and [A2,B2,C2,D2] of the plants P_1 and P_2 respectively, gap calculates g1, the directed gap between P_1 and P_2, and g2, the directed gap between P_2 and P_1, within an accuracy level, given by tol. The gap between P_1 and P_2, which equals the maximum of the two directed gaps, is returned in the variable dist. So in formulae we have:

$$g1 = \vec{\delta}(P_1, P_2),$$

$$g2 = \vec{\delta}(P_2, P_1),$$

$$dist = \delta(P_1, P_2) = \max \{ \vec{\delta}(P_1, P_2), \vec{\delta}(P_2, P_1) \},$$

where we didn't take the accuracy level tol into account.

<u>Algorithm</u> :
gap uses the method given in [6, ch.2] to calculate the gap between two systems. Let (D_i , N_i) and ($\tilde{N}_i$, $\tilde{D}_i$) be normalized right– and left–Bezout factorizations of P_i respectively (i = 1,2). Then, according to (2.17), we have:

$$\vec{\delta}(P_1, P_2) = \inf_{Q \in M(\mathbb{R}H_\infty)} \| \begin{bmatrix} D_1 \\ N_1 \end{bmatrix} - \begin{bmatrix} D_2 \\ N_2 \end{bmatrix} Q \|_\infty. \tag{A.1}$$

When we define

$$G := D_2^* D_1 + N_2^* N_1, \tag{A.2}$$

$$J_1 := -\tilde{N}_2 D_1 + \tilde{D}_2 N_1, \tag{A.3}$$

we can rewrite (A.1) as (see [6, p.5])

$$\vec{\delta}(P_1, P_2) = \inf_{Q \in M(\mathbb{R}H_\infty)} \left\| \begin{bmatrix} G - Q \\ J_1 \end{bmatrix} \right\|. \tag{A.4}$$

Given state–space realizations of the plants P_1 and P_2 it is possible to construct state–space realizations of the matrices G and J_1, as defined in (A.2) and (A.3) (with help of the functions norcor, tfmult and tfadd). Given these realizations of G and J_1, the MATLAB–function dirgap is used to calculate the directed gap $\vec{\delta}(P_1, P_2)$ within the desired level of accuracy.

The calculation of the directed gap between P_2 and P_1 takes place in the same way, where G and J_1 are replaced by G^* and $J_2 := -\tilde{N}_1 D_2 + \tilde{D}_1 N_2$ respectively, i.e.

$$\vec{\delta}(P_2, P_1) = \inf_{Q \in M(\mathbb{R}H_\infty)} \left\| \begin{bmatrix} G^* - Q \\ J_2 \end{bmatrix} \right\|. \tag{A.5}$$

Finally the value of dist, the gap between P_1 and P_2, is calculated as the maximum of the two directed gaps g1 and g2.

WARNING: gap is not suitable to compute gaps between large order systems, certainly not in combination with a small value for tol, the desired accuracy level. The computation becomes then very expensive.

Nested functions :
norcor; see appendix B. dirgap; see appendix A.
tfadd, tfmult; see appendix E.

Program :

```
function [g1,g2,dist] = gap(A1,B1,C1,D1,A2,B2,C2,D2,tol)
```

Given minimal realizations of P1=[A1,B1,C1,D1] and P2=[A2,B2,C2,D2] gap calculates dist, the gap between P1 and P2, where the tolerance for the accuracy is equal to tol. g1 and g2 are the directed gaps between P1 and P2 respectively.

```
[DRA1,DRB1,DRC1,DRD1,NRA1,NRB1,NRC1,NRD1,..
  DLA1,DLB1,DLC1,DLD1,NLA1,NLB1,NLC1,NLD1]=norcor(A1,B1,C1,D1);
```

```
[DRA2,DRB2,DRC2,DRD2,NRA2,NRB2,NRC2,NRD2,..
 DLA2,DLB2,DLC2,DLD2,NLA2,NLB2,NLC2,NLD2]=norcor(A2,B2,C2,D2);


[H1,H2,H3,H4]=tfmult(-DRA2',DRC2',-DRB2',DRD2',DRA1,DRB1,DRC1,DRD1);
[H5,H6,H7,H8]=tfmult(-NRA2',NRC2',-NRB2',NRD2',NRA1,NRB1,NRC1,NRD1);
[GA,GB,GC,GD]=tfadd(H1,H2,H3,H4,H5,H6,H7,H8);
GSA=-GA';GSB=GC';GSC=-GB';GSD=GD';


[H1,H2,H3,H4]=tfmult(NLA2,NLB2,-NLC2,-NLD2,DRA1,DRB1,DRC1,DRD1);
[H5,H6,H7,H8]=tfmult(DLA2,DLB2,DLC2,DLD2,NRA1,NRB1,NRC1,NRD1);
[JA1,JB1,JC1,JD1]=tfadd(H1,H2,H3,H4,H5,H6,H7,H8);


[H1,H2,H3,H4]=tfmult(NLA1,NLB1,-NLC1,-NLD1,DRA2,DRB2,DRC2,DRD2);
[H5,H6,H7,H8]=tfmult(DLA1,DLB1,DLC1,DLD1,NRA2,NRB2,NRC2,NRD2);
[JA2,JB2,JC2,JD2]=tfadd(H1,H2,H3,H4,H5,H6,H7,H8);


g1=dirgap(GA,GB,GC,GD,JA1,JB1,JC1,JD1,tol);
g2=dirgap(GSA,GSB,GSC,GSD,JA2,JB2,JC2,JD2,tol);
dist=max([g1 g2]);
```

dirgap

Purpose :
Computation of the directed gap between two systems. dirgap is a special purpose algorithm,
to be used in the MATLAB–function gap.

Synopsis :
alpha = dirgap(GA,GB,GC,GD,JA,JB,JC,JD,tol)

Description :
Given realizations [GA,GB,GC,GD] of a matrix G, and [JA,JB,JC,JD] of a matrix J, dirgap
computes, under the assumption that

$$\inf_{Q \,\in\, M(\mathbb{R}H_\infty)} \left\| \begin{bmatrix} G - Q \\ J \end{bmatrix} \right\| \leq 1,$$

the value of this infimum (within an accuracy level of tol), and stores it in the variable
alpha:

$$\text{alpha} = \inf_{Q \in M(\mathbb{R}H_\infty)} \left\| \begin{bmatrix} G - Q \\ J \end{bmatrix} \right\|. \tag{A.6}$$

The relation of this formula with the directed gap is already explained in the function–description of the function gap. Note that in the case where G and J are defined as in (A.2) and (A.3), the value of alpha is equal to a directed gap, and so smaller or equal to one. So in this case (the only one we are interested in), the assumption

$$\inf_{Q \in M(\mathbb{R}H_\infty)} \left\| \begin{bmatrix} G - Q \\ J \end{bmatrix} \right\| \leq 1,$$

is always satisfied.

<u>Algorithm</u> :

dirgap is a computer–implementation of the algorithm for the calculation of formula (A.6) as given in [6, p.6], which is based upon the algorithm of Francis described in [4, sec.8.1.]. It proceeds as follows.

From formula (A.6) and our assumption it follows immediately that

$$\| J \| \leq \text{alpha} \leq 1.$$

Now define $\mu := \| J \|$ and suppose $\mu < \gamma < 1$. Let F_γ be a spectral factor of the matrix

$$\gamma^2 I - J^*J,$$

i.e.

$$F_\gamma^* F_\gamma = \gamma^2 I - J^*J, \tag{A.7}$$

with F_γ and $F_\gamma^{-1} \in M(\mathbb{R}H_\infty)$. In [4, sec.7.3.] it is proven that such a factorization really exists. Now it can be shown that (see [4, sec.8.1.]) that

$$\inf_{Q \in M(\mathbb{R}H_\infty)} \left\| \begin{bmatrix} G - Q \\ J \end{bmatrix} \right\| \leq \gamma \quad \Leftrightarrow \quad \| \Gamma_{GF_\gamma^{-1}} \| \leq 1. \tag{A.8}$$

where $\Gamma_{GF_\gamma^{-1}}$ denotes the Hankel–operator with symbol GF_γ^{-1}. Because we know how to calculate the norm of this Hankel–operator (see the MATLAB–function hankel), it is now possible, with help of formula (A.8), to approximate the value of alpha to any wished accuracy by performing an iterative search on the interval $[\mu,1]$. Bisection is used to carry out this search.

WARNING: Remark that in each step of the algorithm a quite expensive spectral factorization has to be carried out. Therefore this method is not very good applicable for matrices of a large order, and certainly not in combination with a small value for tol, the desired level of accuracy.

Nested functions :
hinfn; see [1]; hankel; see appendix C.
tfinv, tfmult; see appendix E.

Program :

function alpha=dirgap(GA,GB,GC,GD,JA,JB,JC,JD,tol)

Given realizations [GA,GB,GC,GD] of G and [JA,JB,JC,JD] of J, dirgap calculates:

$$alpha = \inf_{Q \in M(\mathbb{R}H_\infty)} \left\| \begin{bmatrix} G \vdots Q \\ J \end{bmatrix} \right\|_\infty,$$

within an accuracy of tol. (This under the assumption alpha $\leq$ 1.)

```
[mu,freq,l]=hinfn(JA,JB,JC,JD,1e−8,0);
under=mu;upper=1;

while ((upper−under)>tol)
      gamma=(upper+under)/2;
      [FA,FB,FCh,FDh]=sfl(JA,JB,((1/gamma)*JC),((1/gamma)*JD));
      FC=gamma*FCh;FD=gamma*FDh;
      [FIA,FIB,FIC,FID]=tfinv(FA,FB,FC,FD);
      [RA,RB,RC,RD]=tfmult(GA,GB,GC,GD,FIA,FIB,FIC,FID);
      han=hankel(RA,RB,RC,RD);
      if han ≤ 1
            upper=gamma;
      else
            under=gamma;
      end
end

alpha=(upper+under)/2;
```

APPENDIX B

This appendix contains function–descriptions of the MATLAB–functions ncoprfac, norcor and speccopr.

ncoprfac, norcor

Purpose :
Normalized doubly–Bezout factorization.

Synopsis :
[DRA,DRB,DRC,DRD,NRA,NRB,NRC,NRD,..
  YRA,YRB,YRC,YRD,ZRA,ZRB,ZRC,ZRD,..
  DLA,DLB,DLC,DLD,NLA,NLB,NLC,NLD,..
  YLA,YLB,YLC,YLD,ZLA,ZLB,ZLC,ZLD] = ncoprfac(A,B,C,D)

[DRA,DRB,DRC,DRD,NRA,NRB,NRC,NRD,..
  DLA,DLB,DLC,DLD,NLA,NLB,NLC,NLD] = norcor(A,B,C,D)

Description :
Given a minimal realization [A,B,C,D] of a transfermatrix P(s), ncoprfac computes state–space realizations of a normalized right– and left–Bezout factorization $((D_0,N_0)$ and $(\tilde{N}_0,\tilde{D}_0)$ respectively) and of matrices $Y_0, Z_0, \tilde{Y}_0$ and $\tilde{Z}_0$ such that:

$$\begin{bmatrix} -Z_0 & Y_0 \\ \tilde{D}_0 & \tilde{N}_0 \end{bmatrix} \begin{bmatrix} -N_0 & \tilde{Y}_0 \\ D_0 & \tilde{Z}_0 \end{bmatrix} = \begin{bmatrix} I & 0 \\ 0 & I \end{bmatrix} \tag{B.1}$$

These state–space realizations are given by

$$D_0(s) = [DRA,DRB,DRC,DRD]; \qquad \tilde{D}_0(s) = [DLA,DLB,DLC,DLD];$$

$$N_0(s) = [NRA,NRB,NRC,NRD]; \qquad \tilde{N}_0(s) = [NLA,NLB,NLC,NLD];$$

$$Y_0(s) = [YRA,YRB,YRC,YRD]; \qquad \tilde{Y}_0(s) = [YLA,YLB,YLC,YLD];$$

$$Z_0(s) = [ZRA,ZRB,ZRC,ZRD]; \qquad \tilde{Z}_0(s) = [ZLA,ZLB,ZLC,ZLD].$$

norcor is an abbreviated version of ncoprfac and produces only state–space realizations of the normalized right– and left–Bezout factorizations (D_0,N_0) and $(\tilde{N}_0,\tilde{D}_0)$ of P.

Algorithm :

ncoprfac and norcor use the construction method for state–space realizations of normalized Bezout factorizations as given in theorem 4.1.1.

Nested functions :

sqrtl; see appendix E.

See also :

speccopr (see appendix B); this is a special purpose algorithm to be used in other algorithms.

Programs :

```
function    [DRA,DRB,DRC,DRD,NRA,NRB,NRC,NRD,..
            YRA,YRB,YRC,YRD,ZRA,ZRB,ZRC,ZRD,..
            DLA,DLB,DLC,DLD,NLA,NLB,NLC,NLD,..
            YLA,YLB,YLC,YLD,ZLA,ZLB,ZLC,ZLD]=ncoprfac(A,B,C,D)
```

Given a minimal realization [A,B,C,D] of a transfermatrix P(s), ncoprfac computes realizations of a normalized right coprime factorization of P. P=ND^{-1} with D=[DRA,DRB,DRC,DRD] and N=[NRA,NRB, NRC,NRD]; Y=[YRA,YRB,YRC,YRD] and Z=[ZRA,ZRB,ZRC,ZRD] are such that YD+ZN=I. The left coprime case is treated in a similar way.

```
[m,n]=size(D);
H=eye(n)+(D'*D);L=eye(m)+(D*D');
Hinv=inv(H);Linv=inv(L);
Ahulp1=A–(B*Hinv*D'*C);
Ahulp2=(A–(B*D'*Linv*C))';

[X,XPERR]=aresolv(Ahulp1,C'*Linv*C,B*Hinv*B','Schur');
[Y,YPERR]=aresolv(Ahulp2,B*Hinv*B',C'*Linv*C,'Schur');

F=Hinv*((D'*C)+(B'*X));  AC=A–(B*F);
K=((B*D')+(Y*C'))*Linv;  AO=A–(K*C);

Hr=sqrtl(H);Hrinv=inv(Hr);
Lr=sqrtl(L);Lrinv=inv(Lr);

DRA=AC;DRB=–B*Hrinv;DRC=F;DRD=Hrinv;
NRA=AC;NRB=B*Hrinv;NRC=C–(D*F);NRD=D*Hrinv;
YRA=AO;YRB=B–(K*D);YRC=Hr*F;YRD=Hr;
```

```
ZRA=AO;ZRB=K;ZRC=Hr*F;ZRD=zeros(n,m);


DLA=AO;DLB=K;DLC=-Lrinv*C;DLD=Lrinv;
NLA=AO;NLB=B-(K*D);NLC=Lrinv*C;NLD=Lrinv*D;
YLA=AC;YLB=K*Lr;YLC=C-(D*F);YLD=Lr;
ZLA=AC;ZLB=K*Lr;ZLC=F;ZLD=zeros(n,m);
```

function   [DRA,DRB,DRC,DRD,NRA,NRB,NRC,NRD,..
                DLA,DLB,DLC,DLD,NLA,NLB,NLC,NLD]=norcor(A,B,C,D)

Given a minimal realization [A,B,C,D] of a transfermatrix P(s), norcor computes realizations of a normalized right coprime factorization of P, (P=ND^{-1} with D=[DRA,DRB,DRC,DRD] and N=[NRA,NRB,NRC,NRD]) and a normalized left coprime factorization of P (P=D^{-1}N with D=[DLA,DLB,DLC,DLD] and N=[NLA,NLB,NLC,NLD]).

```
[m,n]=size(D);
H=eye(n)+(D'*D);L=eye(m)+(D*D');
Hinv=inv(H);Linv=inv(L);
Ahulp1=A-(B*Hinv*D'*C);
Ahulp2=(A-(B*D'*Linv*C))';

[X,XPERR]=aresolv(Ahulp1,C'*Linv*C,B*Hinv*B','Schur');
[Y,YPERR]=aresolv(Ahulp2,B*Hinv*B',C'*Linv*C,'Schur');

F=Hinv*((D'*C)+(B'*X));  AC=A-(B*F);
K=((B*D')+(Y*C'))*Linv;  AO=A-(K*C);

Hr=sqrtl(H);Hrinv=inv(Hr);
Lr=sqrtl(L);Lrinv=inv(Lr);

DRA=AC;DRB=-B*Hrinv;DRC=F;DRD=Hrinv;
NRA=AC;NRB=B*Hrinv;NRC=C-(D*F);NRD=D*Hrinv;

DLA=AO;DLB=K;DLC=-Lrinv*C;DLD=Lrinv;
NLA=AO;NLB=B-(K*D);NLC=Lrinv*C;NLD=Lrinv*D;
```

speccopr

Purpose :
Normalized Bezout factorization of a plant P, and calculation of the matrix V, associated with P. speccopr is a special purpose algorithm, to be used in the functions robstab and robstgl.

Synopsis :
[DRA,DRB,DRC,DRD,NRA,NRB,NRC,NRD,..
  YLA,YLB,YLC,YLD,ZLA,ZLB,ZLC,ZLD,..
  VA,VB,VC,VD] = speccopr(A,B,C,D)

Description :
Given a minimal realization [A,B,C,D] of a transfermatrix P(s), speccopr computes state–space realizations of a normalized right–Bezout factorization (D_0, N_0) of P and of matrices $\tilde{Y}_0$ and $\tilde{Z}_0$ such that (B.1) holds. (So speccopr delivers only a part of the output of ncoprfac.) Also a minimal state–space realization of the matrix V, associated with P as defined in (4.57), is given. These state–space realizations are returned as follows:

$$D_0(s) = [DRA,DRB,DRC,DRD]; \qquad \tilde{Y}_0(s) = [YLA,YLB,YLC,YLD];$$
$$N_0(s) = [NRA,NRB,NRC,NRD]; \qquad \tilde{Z}_0(s) = [ZLA,ZLB,ZLC,ZLD];$$
$$V(s) = [VA,VB,VC,VD].$$

Algorithm :
speccopr uses the construction method of theorem 4.1.1. for the state–space realizations of the matrices $D_0, N_0, \tilde{Y}_0$ and $\tilde{Z}_0$, and the method of theorem 5.2.1. for the state–space realization of V (formula (5.4)).

Nested functions :
sqrtl; see appendix E.

See also :
ncoprfac (see appendix B).

Program :
See next page.

```
function    [DRA,DRB,DRC,DRD,NRA,NRB,NRC,NRD,..
            YLA,YLB,YLC,YLD,ZLA,ZLB,ZLC,ZLD,..
            VA,VB,VC,VD]=speccopr(A,B,C,D)
```

Given a minimal realization [A,B,C,D] of a transfermatrix P, there are an n.c.r.f. (D,N) of P and an n.l.c.f.

(D̄,N̄) of P and stable matrices Y, Z and Ȳ, Z̄, such that

$$\begin{bmatrix} -Z & Y \\ \bar{D} & \bar{N} \end{bmatrix} \begin{bmatrix} -N & \bar{Y} \\ D & \bar{Z} \end{bmatrix} = \begin{bmatrix} I & 0 \\ 0 & I \end{bmatrix}.$$

speccopr computes state space realizations of D=[DRA,DRB,DRC,DRD], N=[NRA,NRB,NRC,NRD],

Ȳ=[YLA,YLB,YLC,YLD] and Z̄=[ZLA,ZLB,ZLC,ZLD]. Moreover, a state space realization of the matrix

V = D*Z̄ - N*Ȳ (V=[VA,VB,VC,VD]) is computed.

```
[m,n]=size(D);[p,q]=size(A);
H=eye(n)+(D'*D);L=eye(m)+(D*D');
Hinv=inv(H);Linv=inv(L);
Ahulp1=A-(B*Hinv*D'*C);
Ahulp2=(A-(B*D'*Linv*C))';

[X,XPERR]=aresolv(Ahulp1,C'*Linv*C,B*Hinv*B','Schur');
[Y,YPERR]=aresolv(Ahulp2,B*Hinv*B',C'*Linv*C,'Schur');

F=Hinv*((D'*C)+(B'*X));  AC=A-(B*F);
K=((B*D')+(Y*C'))*Linv;  AO=A-(K*C);

Hr=sqrtl(H);Hrinv=inv(Hr);
Lr=sqrtl(L);Lrinv=inv(Lr);

DRA=AC;DRB=-B*Hrinv;DRC=F;DRD=Hrinv;
NRA=AC;NRB=B*Hrinv;NRC=C-(D*F);NRD=D*Hrinv;
YLA=AC;YLB=K*Lr;YLC=C-(D*F);YLD=Lr;
ZLA=AC;ZLB=K*Lr;ZLC=F;ZLD=zeros(n,m);

VA=-AC';VB=(eye(p)+(X*Y))*C'*Lrinv;
VC=Hrinv*B';VD=-Hrinv*D'*Lr;
```

APPENDIX C

This appendix contains function–descriptions of the MATLAB–functions hankel, hanneh and robstab.

hankel

<u>Purpose</u> :
Computation of the norm of a Hankel–operator.

<u>Synopsis</u> :
norm = hankel(A,B,C,D)

<u>Description</u> :
Given a (not necessary minimal) realization [A,B,C,D] of a real–rational matrix V(s), hankel computes the norm of the Hankel–operator with symbol V.

<u>Algorithm</u> :
hankel uses the algorithm for the computation of the norm of a Hankel–operator as given in section 5.3.

<u>See also</u> :
hanneh (see appendix C); this function not only computes the norm of the Hankel–operator with symbol V, but also solves the related Nehari–problem.

<u>Program</u> :

function norm=hankel(A,B,C,D)

Given a realization [A,B,C,D] of a real rational matrix V(s) hankel computes the norm of the hankel-operator with symbol V.

```
[A1,B1,C1,D1,A4,B2,C2,D2,m]=stabproj(A,B,C,D);
BBT2=B2*(B2');CTC2=(C2')*C2;
LC=lyap(A4,A4',-BBT2);LO=lyap(A4',A4,-CTC2);
LCLO=LC*LO;eigenval=eig(LCLO);a=max(eigenval);norm=sqrt(a);
```

hanneh

<u>Purpose</u> :
Computation of the norm of a Hankel–operator and of a sub–optimal solution to the related Nehari–problem.

<u>Synopsis</u> :
[XA,XB,XC,XD,hannorm] = hanneh(A,B,C,D,tol)

<u>Description</u> :
Given a minimal realization [A,B,C,D] of a real–rational matrix V(s), hanneh computes:
i)   hannorm; the norm of the Hankel–operator with symbol V.
ii)   a realization [XA,XB,XC,XD] of a matrix X(s) $\in$ M($\mathbb{R}$H$_\infty$) which is a sub–optimal solution to the related Nehari–problem, within the desired accuracy–level, given by 'tol'. I.e. a matrix X(s) $\in$ M($\mathbb{R}$H$_\infty$) is calculated such that:

$$\| \ V - X \ \| \ < \ \| \ \Gamma_V \ \| \ (1 + \text{tol} \ ) \qquad\qquad (C.1)$$

<u>Algorithm</u> :
hanneh uses the second algorithm stated in section 5.4. for the solution of the Nehari–problem. This algorithm is based upon the method for solving the Nehari–problem described in the same paragraph (this is the same method, Francis gives in [4, ch.5]). In this method the norm of the Hankel–operator is calculated automatically, actually it is an intermediate result.

<u>See also</u> :
hannehgl; this function gives for an anti–stable matrix V(s) an exact solution to the Nehari–problem (see appendix D).

<u>Program</u> :

function[XA,XB,XC,XD,hannorm]=hanneh(A,B,C,D,tol)

Given a minimal realization [A,B,C,D] of a real rational matrix V(s) hanneh computes:
i)   hannorm: the norm of the hankel-operator with symbol V
ii)   a realization [XA,XB,XC,XD] of a matrix X(s) such that the Hinf-norm of V-X smaller is than hannorm*(1+tol)

[A1,B1,C1,D1,A4,B2,C2,D2,m]=stabproj(A,B,C,D);
BBT2=B2*(B2');CTC2=(C2')*C2;

```
LC=lyap(A4,A4,–BBT2);
LO=lyap(A4',A4,–CTC2);
LCLO=LC*LO;eigenval=eig(LCLO);a=max(eigenval);

a1=1/(a*(1+tol)*(1+tol));
[p,q]=size(LCLO);
Ninv=eye(p)–(a1*LCLO);N=inv(Ninv);

[r,s]=size(A);[k,l]=size(C);
hulp1=a1*N*LO*B2;hulp2=(–N')*BBT2;

XA=[A zeros(r,(2*p));zeros(p,r) A4 hulp2;zeros(p,(r+p)) (–A4'–(hulp1*B2'))];
XB=[B;N'*B2;hulp1];
XC=[C –C2 zeros(k,p)];
XD=D;

hannorm=sqrt(a);
```

robstab

Purpose :

Computation of the maximal stability radius belonging to a plant P and a sub–optimally robust controller C.

Synopsis :

[CA,CB,CC,CD,bound] = robstab(PA,PB,PC,PD,tol)

Description :

Given a minimal realization [PA,PB,PC,PD] of a plant P, robstab computes the maximal stability radius belonging to P (this value is stored in the variable bound). Also a state–space realization [CA,CB,CC,CD] of a sub–optimally robust controller C is calculated, which realizes the value of bound within the desired accuracy–level, given by tol. So not only P is stabilized by C, but also the systems in the ball around P, with radius (bound)*(1–tol), i.e. the whole set

$$\{ \ P_\lambda \in M(\mathbb{R}(s)) \ | \ \delta(P,P_\lambda) < (bound) * (1 - tol) \ \} \tag{C.2}$$

is stabilized by C.

When the order of the original plant P is n, the order of the calculated compensator is 10n.

Algorithm :

robstab is the computer–implementation of the algorithm for the solution of the problem of (sub)–optimally robust control as given in section 5.5. This means that the internal Nehari–problem is solved by the method of section 5.4., which is implemented in the function hanneh.

See also :

robstgl, rbstglfar (see appendix D); these are other algorithms for the solution of the problem of optimally robust control.

Nested functions :

speccopr; see appendix B. hanneh; see appendix C.

tfadd, tfinv, tfmult; see appendix E.

Program :

```
function[CA,CB,CC,CD,bound]=robstab(PA,PB,PC,PD,tol)
```

Given plant P=[PA,PB,PC,PD], robstab calculates a compensator C=[CA,CB,CC,CD] that stabilizes P, and for each plant Q, such that d(P,Q) < bound*(1-tol), C is also a stabilizing compensator. The variable bound represents the maximal stability radius belonging to P, i.e. the radius of the largest ball around P that can be stabilized by one compensator. In this way C is a sub-optimal solution to the problem of optimally robust control.

In the computation a Nehari-problem has to be solved. tol determines how accurate this problem has to be solved (tolerance in accuracy).

```
[DRA,DRB,DRC,DRD,NRA,NRB,NRC,NRD,YLA,YLB,..
  YLC,YLD,ZLA,ZLB,ZLC,ZLD,VA,VB,VC,VD]=speccopr(PA,PB,PC,PD);
[RA,RB,RC,RD,Ahan]=hanneh(VA,VB,VC,VD,tol);
wg=sqrt(1+(Ahan*Ahan));  bound=1/wg;


[H1,H2,H3,H4]=tfmult(DRA,DRB,DRC,DRD,RA,RB,-RC,-RD);
[H5,H6,H7,H8]=tfadd(ZLA,ZLB,ZLC,ZLD,H1,H2,H3,H4);


[H1,H2,H3,H4]=tfmult(NRA,NRB,NRC,NRD,RA,RB,RC,RD);
[H9,H10,H11,H12]=tfadd(YLA,YLB,YLC,YLD,H1,H2,H3,H4);
[H1,H2,H3,H4]=tfinv(H9,H10,H11,H12);


[CA,CB,CC,CD]=tfmult(H5,H6,H7,H8,H1,H2,H3,H4);
```

APPENDIX D

This appendix contains function–descriptions of the MATLAB–functions hannehgl, robstgl and rbstglfar.

hannehgl

Purpose :
Computation of the norm of a Hankel–operator, and of an exact solution to the related Nehari–problem.

Synopsis :
[XA,XB,XC,XD,hannorm] = hannehgl(A,B,C,D)

Description :
Given a minimal state–space realization [A,B,C,D] of a real–rational *anti–stable* matrix V(s), hannehgl computes:

i)   hannorm, the norm of the Hankel–operator with symbol V.

ii)   a realization [XA,XB,XC,XD] of a matrix $X(s) \in M(\mathbb{R}H_\infty)$, which is an exact solution to the Nehari–problem for the matrix V(s). I.e. the matrix $X(s) \in M(\mathbb{R}H_\infty)$ has the property that

$$\| V - X \| = \| \Gamma_V \|. \tag{D.1}$$

When the order of the matrix V is n, the state–space realization of the matrix X is of the order n–1.

Algorithm :
hannehgl uses the method of Glover as described in section 6.2. to solve the Nehari–problem exactly. The realization of X is constructed without the use of balanced realizations, but with the method given in [14] and implemented in the MATLAB–function ohkapp. This last function is a standard function in the Robust Control Toolbox.

See also :
hanneh (see appendix B); this algorithm computes a sub–optimal solution to the Nehari–problem.

Program :
See next page.

function[XA,XB,XC,XD,hannorm]=hannehgl(A,B,C,D)

Given a minimal realization [A,B,C,D] of a real anti-stable matrix V(s), hannehgl computes:

i)   *hannorm: the norm of the Hankel-operator with symbol V*

ii)   *a realization [XA,XB,XC,XD] of a stable matrix X(s), such that the Hinf-norm of V-X is equal to*
         hannorm, and the order of X is smaller than the order of V

[AX,BX,CX,DX,AY,BY,CY,DY,aug]=ohkapp(−A,B,−C,D,1,0);
XA=−AY;XB=BY;XC=−CY;XD=DY;
hannorm=aug(1,1);

robstgl

Purpose :
Computation of the maximal stability radius belonging to a plant P and an optimally robust controller C.

Synopsis :
[CA,CB,CC,CD,bound] = robstgl(PA,PB,PC,PD)

Description :
Given a minimal realization [PA,PB,PC,PD] of a plant P, robstgl computes the maximal stability radius belonging to P (stored in the variable bound), and a state–space realization [CA,CB,CC,CD] of an optimally robust controller C that realizes this bound. This controller C not only stabilizes P, but also systems in the ball around P, with radius 'bound', i.e. the set

$$\{ \ P_\lambda \in M(\mathbb{R}(s)) \ | \ \delta(P,P_\lambda) < \text{bound} \ \}. \hspace{2cm} (D.2)$$

When the order of the plant is n, the order of the calculated compensator C is 6n−2.

Algorithm :
robstgl uses the method of Glover for the solution of the problem of optimally robust control, as explained in section 6.2. So the internal Nehari–problem is exactly solved by the method of Glover, which is implemented in the function hannehgl.

See also :
robstab (see appendix C), rbstglfar (see appendix D); these are other algorithms for the solution of the problem of optimally robust control.

Nested functions :
speccopr; see appendix B. hannehgl; see appendix D.
tfadd, tfinv, tfmult; see appendix E.

Program :

function[CA,CB,CC,CD,bound]=robstgl(PA,PB,PC,PD)

Given plant P=[PA,PB,PC,PD], robstgl calculates a compensator C=[CA,CB,CC,CD] that stabilizes P, and for each plant Q, such that d(P,Q)<bound, C is also a stabilizing compensator. Moreover, C is chosen in such a way that bound is maximal. I.e. for each compensator C' that stabilizes P, the radius of the ball around P of plants also stabilized by C', is not greater than bound.

[DRA,DRB,DRC,DRD,NRA,NRB,NRC,NRD,YLA,YLB,..
  YLC,YLD,ZLA,ZLB,ZLC,ZLD,VA,VB,VC,VD]=speccopr(PA,PB,PC,PD);
[RA,RB,RC,RD,Ahan]=hannehgl(VA,VB,VC,VD);
wg=sqrt(1+(Ahan*Ahan)); bound=1/wg;

[H1,H2,H3,H4]=tfmult(DRA,DRB,DRC,DRD,RA,RB,-RC,-RD);
[H5,H6,H7,H8]=tfadd(ZLA,ZLB,ZLC,ZLD,H1,H2,H3,H4);
[H1,H2,H3,H4]=tfmult(NRA,NRB,NRC,NRD,RA,RB,RC,RD);
[H9,H10,H11,H12]=tfadd(YLA,YLB,YLC,YLD,H1,H2,H3,H4);
[H1,H2,H3,H4]=tfinv(H9,H10,H11,H12);

[CA,CB,CC,CD]=tfmult(H5,H6,H7,H8,H1,H2,H3,H4);

rbstglfar

Purpose :
Computation of the maximal stability radius belonging to a plant P and a sub–optimally robust controller C.

Synopsis :
[CA,CB,CC,CD,bound] = rbstglfar(PA,PB,PC,PD,tol)

Description :
Given a minimal realization [PA,PB,PC,PD] of a plant P, rbstglfar computes the maximal stability radius belonging to P (which is returned in the variable bound) and a state–space

realization [CA,CB,CC,CD] of a sub–optimally robust controller C, which realizes this bound within an accuracy–level of tol (tol is one of the input variables). This means that C stabilizes all the systems in the ball around P with radius (bound)*(1–tol), i.e. C stabilizes the set

$$\{\ P_\lambda\ \epsilon\ M(\mathbb{R}(s))\ |\ \delta(P,P_\lambda) < (bound) * (1-tol)\ \}. \tag{D.3}$$

The order of this compensator C is equal to the order of the original plant P.

Algorithm :

To compute the maximal stability radius belonging to P, rbstglfar uses the method of theorem 6.3.2. So the value of bound is calculated with formula (6.22). The state–space realization of C is constructed by the method of Glover and McFarlane as given in theorem 6.3.3. Formula (6.35) is used for the computation of the realization of C.

See also :

robstab (see appendix C), robstgl (see appendix D); these are other algorithms for the solution of the problem of optimally robust control.

Program :

```
function[CA,CB,CC,CD,bound]=rbstglfar(PA,PB,PC,PD,tol)
```

Given a minimal realization [PA,PB,PC,PD] of a plant P, rbstglfar calculates the maximal stability radius 'bound' and a realization [CA,CB,CC,CD] of a compensator C, which realizes this bound within an accuracy-level of tol.

```
[m,n]=size(PD);[r,k]=size(PA);
H=eye(n)+(PD'*PD);L=eye(m)+(PD*PD');
Hinv=inv(H);Linv=inv(L);
Ahulp1=PA–(PB*Hinv*PD'*PC);
Ahulp2=(PA–(PB*PD'*Linv*PC))';

[X,XPERR]=aresolv(Ahulp1,PC'*Linv*PC,PB*Hinv*PB','Schur');
[Y,YPERR]=aresolv(Ahulp2,PB*Hinv*PB',PC'*Linv*PC,'Schur');

l=max(eig(Y*X));bound=1/sqrt(1+l);
gamma=(1/bound)*(1+tol);g2=gamma*gamma;

F=Hinv*((PD'*PC)+(PB'*X));
```

```
AC=PA–(PB*F);
W1=((1–g2)*eye(r))+(X*Y);

CB=g2*(inv(W1'))*Y*PC';
CA=AC+CB*(PC–(PD*F));
CC=PB'*X;
CD=–PD';
```

APPENDIX E

This appendix contains function–descriptions of the MATLAB–functions tfadd, tfinv, tfmult and sqrtl.

tfadd

Purpose :
Addition of two transfermatrices in state–space form.

Synopsis :
[P,Q,R,S] = tfadd(A1,B1,C1,D1,A2,B2,C2,D2)

Description :
Given state–space realizations of two transfermatrices of the same size, tfadd computes a state–space realization of the sum of these transfermatrices, i.e.

$$[P,Q,R,S] = [A1,B1,C1,D1] + [A2,B2,C2,D2].$$

Algorithm :
tfadd uses the state–space formula of page ix for the addition of two transfermatrices, to construct a state–space realization of the sum.

Program :

```
function[P,Q,R,S]=tfadd(A1,B1,C1,D1,A2,B2,C2,D2)
```

Given state space realizations of two transfermatrices of the same size, tfadd calculates a state space realization of the sum of these transfermatrices, i.e.
$$[P,Q,R,S]=[A1,B1,C1,D1]+[A2,B2,C2,D2].$$

```
[na1,ma1]=size(A1);  [na2,ma2]=size(A2);
P=[A1 zeros(na1,ma2);zeros(na2,ma1) A2];
Q=[B1;B2];
R=[C1 C2];
S=D1+D2;
```

tfinv

Purpose :
Inversion of a transfermatrix in state–space form.

Synopsis :
[P,Q,R,S] = tfinv(A,B,C,D)

Description :
Given a state–space realization of a square invertible transfermatrix, tfinv computes a state–space realization of the inverse of this transfermatrix, i.e.

$$[P,Q,R,S] = [A,B,C,D]^{-1}.$$

Algorithm :
tfinv uses the state–space formula of page ix for the inversion of a transfermatrix, to construct a state–space realization of the inverse.

Program :

```
function[P,Q,R,S]=tfinv(A,B,C,D)
```

Given a state space realization of a square transfermatrix (with D invertible), tfinv calculates a state space realization of the inverse of this transfermatrix. So
$$[P,Q,R,S]=inv([A,B,C,D]).$$

```
hulp=inv(D);
P=A–(B*hulp*C);
Q=B*hulp;
R=–hulp*C;
S=hulp;
```

tfmult

<u>Purpose</u> :
Multiplication of two transfermatrices in state–space form.

<u>Synopsis</u> :
[P,Q,R,S] = tfmult(A1,B1,C1,D1,A2,B2,C2,D2)

<u>Description</u> :
Given state–space realizations of two transfermatrices of compatible size (the number of
columns of the first one equals the number of rows of the second one), tfmult calculates a
state–space realization of the product of these transfermatrices, i.e.

$$[P,Q,R,S] = [A1,B1,C1,D1] * [A2,B2,C2,D2].$$

<u>Algorithm</u> :
tfmult uses the state–space formula of page ix for the multiplication of two transfermatrices,
to construct a state–space realization of the product.

<u>Program</u> :

```
function[P,Q,R,S]=tfmult(A1,B1,C1,D1,A2,B2,C2,D2)
```

*Given state space realizations of two transfermatrices (number of columns of the first equals the number of
rows of the second), tfmult computes a state space realization of the product of the two transfermatrices,
i.e.*
[P,Q,R,S]=[A1,B1,C1,D1]*[A2,B2,C2,D2].

```
[na1,ma1]=size(A1);  [na2,ma2]=size(A2);
P=[A1  (B1*C2);zeros(na2,ma1)  A2];
Q=[(B1*D2);B2];
R=[C1  (D1*C2)];
S=D1*D2;
```

sqrtl

Purpose :

Computation of the square root of a positive definite matrix.

Synopsis :

root = sqrtl(A)

Description :

Given a positive definite matrix A, sqrtl computes the matrix $A^{1/2}$ with help of the singular value decomposition. So when A is given by

$$A = U \ S \ V^T,$$

with S diagonal, then sqrtl(A) is

$$sqrtl(A) = U \ S^{1/2} \ V^T.$$

Program :

function wortel=sqrtl(A)

sqrtl computes the square root of a PD matrix with help of the SV-Decomposition.

```
[U,S,V]=svd(A);
D=S.^(1/2);
wortel=U*D*V;
```

APPENDIX F

This appendix contains solutions to the problem of (sub)–optimally robust stabilization for a plant with transferfunction $P(s) = \frac{1}{s-1}$, calculated in five different ways:

i)   with the MATLAB–function robstab; tolerance level tol = 10^{-5};
ii)   with the MATLAB–function robstab; tolerance level tol = 10^{-8};
iii)   with the MATLAB–function robstgl;
iv)   with the MATLAB–function rbstglfar; tolerance level tol = 10^{-5};
v)   with the MATLAB–function rbstglfar; tolerance level tol = 10^{-8}.

In each case the calculated maximal stability radius is given and a state–space realization of a compensator that realizes this bound within (for the cases i), ii), iv) and v)) the desired accuracy level.

i) MATLAB–function robstab, tolerance level tol = 10^{-5}

PA = 1

PB = 1

PC = 1

PD = 0

bound = 0.3827

tol = 1.0000e–005

CA =   1.0e+006 *

Columns 1 through 7

```
 -0.0000        0        0        0        0   -0.0000   -0.0000
       0  -0.0000   0.0000  -0.0000        0        0         0
       0        0   0.0000        0        0   -0.0000   -0.0000
       0        0        0   0.0000  -2.3314  -0.3414   -0.3414
       0        0        0        0  -0.1414  -0.0207   -0.0207
       0        0        0        0        0   -0.0000   -0.0000
       0        0        0        0        0        0   -0.0000
       0        0        0        0        0   -0.0000   -0.0000
       0        0        0        0        0   -0.3414   -0.3414
       0        0        0        0        0   -0.0207   -0.0207
```

Columns 8 through 10

```
       0        0        0
       0        0        0
       0        0        0
       0        0        0
       0        0        0
       0        0        0
  0.0000  -0.0000        0
  0.0000        0        0
       0   0.0000  -2.3314
       0        0  -0.1414
```

CB =   1.0e+005 *

```
  0.0000
       0
  0.0001
  3.4143
  0.2071
  0.0000
       0
  0.0001
  3.4143
  0.2071
```

CC =

Columns 1 through 7

```
  2.4142    2.4142   -1.0000    1.0000         0         0         0
```

Columns 8 through 10

```
       0         0         0
```

CD = 0

ii) MATLAB—function robstab, tolerance level tol = 10^{-8}

PA = 1

PB = 1

PC = 1

PD = 0

bound = 0.3827

tol = 1.0000e−008

CA =   1.0e+009 *

Columns 1 through 7

−0.0000	0	0	0	0	−0.0000	−0.0000
0	−0.0000	0.0000	−0.0000	0	0	0
0	0	0.0000	0	0	−0.0000	−0.0000
0	0	0	0.0000	−2.3314	−0.3414	−0.3414
0	0	0	0	−0.1414	−0.0207	−0.0207
0	0	0	0	0	−0.0000	−0.0000
0	0	0	0	0	0	−0.0000
0	0	0	0	0	−0.0000	−0.0000
0	0	0	0	0	−0.3414	−0.3414
0	0	0	0	0	−0.0207	−0.0207

Columns 8 through 10

0	0	0
0	0	0
0	0	0
0	0	0
0	0	0
0	0	0
0.0000	−0.0000	0
0.0000	0	0
0	0.0000	−2.3314
0	0	−0.1414

CB =   1.0e+008 *

0.0000
0
0.0000
3.4142
0.2071
0.0000
0
0.0000
3.4142
0.2071

CC =

Columns 1 through 7

    2.4142   2.4142   −1.0000   1.0000   0   0   0

Columns 8 through 10

        0   0   0

CD = 0

iii) MATLAB–function robstgl

PA = 1

PB = 1

PC = 1

PD = 0

bound = 0.3827

CA =

$$
\begin{array}{rrrr}
-1.4142 & 0 & -2.4142 & -2.4142 \\
0 & -1.4142 & 2.4142 & 2.4142 \\
0 & 0 & -3.8284 & -2.4142 \\
0 & 0 & 2.4142 & 1.0000
\end{array}
$$

CB =

$$
\begin{array}{r}
2.4142 \\
-2.4142 \\
2.4142 \\
-2.4142
\end{array}
$$

CC =

    2.4142   2.4142   −2.4142   −2.4142

CD = 2.4142

iv) MATLAB–function rbstglfar, tolerance level tol = 10^{-5}

PA = 1

PB = 1

PC = 1

PD = 0

bound = 0.3827

tol = 1.0000e–005

CA = –1.2071e+005

CB = –1.2071e+005

CC = 2.4142

CD = 0

v) MATLAB–function rbstglfar, tolerance level tol = 10^{-8}

PA = 1

PB = 1

PC = 1

PD = 0

bound = 0.3827

tol = 1.0000e–008

CA = –1.2071e+008

CB = –1.2071e+008

CC = 2.4142

CD = 0

REFERENCES

[1]   N.A. Bruinsma and M. Steinbuch, *A fast algorithm to compute the H_∞-norm of a transferfunction.* Philips Research Laboratories Eindhoven. Eindhoven, 1989.

[2]   R. Chiang and M. Safonov, *Robust-Control Toolbox, User's Guide.* The MathWorks, Inc. South Natick, 1988.

[3]   H.O. Cordes and J.P. Labrousse, The invariance of the index in the metric space of closed operators. *Journal of Mathematics and Mechanics*, vol. 12, pp. 693–719, 1963.

[4]   B.A. Francis, *A Course in H_∞ Control Theory.* Berlin–Heidelberg–New York, Springer Verlag, 1987. Lecture Notes in Control and Information Sciences, vol. 88.

[5]   T.T. Georgiou, On the computation of the gap metric. *Systems and Control Letters*, vol. 11, pp. 253–257, 1988.

[6]   T.T. Georgiou and M.C. Smith, *Optimal Robustness in the Gap Metric.* Preprint, 1989.

[7]   K. Glover, All optimal Hankel–norm approximations of linear multivariable systems and their L_∞–error bounds. *Int. J. Control*, vol. 39, pp. 1115–1193, 1984.

[8]   K. Glover and D. McFarlane, Robust stabilization of normalized coprime factor plant descriptions with H_∞–bounded uncertainty. *IEEE Trans. on Automat. Control*, vol. 34, pp. 821–830, 1989.

[9]   T. Kato, *Perturbation Theory for Linear Operators.* Berlin–New York, Springer Verlag, 1966.

[10]   M.A. Krasnosel'skii, G.M. Vainikko and P.P. Zabreiko, *Approximate Solution to Operator Equations.* Groningen, Wolters–Noordhoff, 1972.

[11]   E. Kreyszig, *Introductory Functional Analysis with Applications.* New York, Wiley, 1978.

[12]   D. Meyer and G. Franklin, A connection between normalized coprime factorizations and linear quadratic regulator theory. *IEEE Trans. on Automat. Control*, vol. AC–32, pp. 227–228, 1987.

[13]   Z. Nehari, On bounded bilinear forms. *Annals of Mathematics*, vol. 65, pp. 153–162, 1957.

[14]   M. Safonov, R. Chiang and D. Limebeer, Hankel model reduction without balancing – A descriptor approach. *Proc. of the 26th Conf. on Decision and Control*, Los Angeles, 1987.

[15]   M. Vidyasagar, *Control System Synthesis: A Factorization Approach.* Cambridge, MA, M.I.T. Press, 1985.

[16]   M. Vidyasagar, Normalized coprime factorizations for nonstrictly proper systems. *IEEE Trans. on Automat. Control*, vol. 33, pp. 300–301, 1988.

[17]   M. Vidyasagar and H. Kimura, Robust controllers for uncertain linear multivariable systems. *Automatica*, vol. 22, pp. 85–94, 1986.

[18]   S.Q. Zhu, Graph topology and gap topology for unstable plants. *IEEE Trans. on Automat. Control*, vol. 34, pp. 848–855, 1989.

[19] S.Q. Zhu, *Robustness of Feedback Stabilization: A topological approach.* Ph. D. dissertation, Eindhoven University of Technology, Eindhoven, 1989.

[20] S.Q. Zhu, M.L.J. Hautus and C. Praagman, Sufficient conditions for robust BIBO stabilization: Given by the gap metric. *Systems and Control Letters*, vol. 11, pp. 53–59, 1988.

Lecture Notes in Control and Information Sciences

Edited by M. Thoma and A. Wyner

Lecture Notes in Control and Information Sciences

Edited by M. Thoma and A. Wyner

Vol. 117: K.J. Hunt
Stochastic Optimal Control Theory
with Application in Self-Tuning Control
X, 308 pages, 1989.

Vol. 118: L. Dai
Singular Control Systems
IX, 332 pages, 1989

Vol. 119: T. Başar, P. Bernhard
Differential Games and Applications
VII, 201 pages, 1989

Vol. 120: L. Trave, A. Titli, A. M. Tarras
Large Scale Systems:
Decentralization, Structure Constraints
and Fixed Modes
XIV, 384 pages, 1989

Vol. 121: A. Blaquière (Editor)
Modeling and Control of Systems
in Engineering, Quantum Mechanics,
Economics and Biosciences
Proceedings of the Bellman Continuum
Workshop 1988, June 13–14, Sophia Antipolis, France
XXVI, 519 pages, 1989

Vol. 122: J. Descusse, M. Fliess, A. Isidori,
D. Leborgne (Eds.)
New Trends in Nonlinear Control Theory
Proceedings of an International
Conference on Nonlinear Systems,
Nantes, France, June 13–17, 1988
VIII, 528 pages, 1989

Vol. 123: C. W. de Silva, A. G. J. MacFarlane
Knowledge-Based Control with
Application to Robots
X, 196 pages, 1989

Vol. 124: A. A. Bahnasawi, M. S. Mahmoud
Control of Partially-Known
Dynamical Systems
XI, 228 pages, 1989

Vol. 125: J. Simon (Ed.)
Control of Boundaries and Stabilization
Proceedings of the IFIP WG 7.2 Conference
Clermont Ferrand, France, June 20–23, 1988
IX, 266 pages, 1989

Vol. 126: N. Christopeit, K. Helmes
M. Kohlmann (Eds.)
Stochastic Differential Systems
Proceedings of the 4th Bad Honnef Conference
June 20–24, 1988
IX, 342 pages, 1989

Vol.127: C. Heij
Deterministic Identification
of Dynamical Systems
VI, 292 pages, 1989

Vol. 128: G. Einarsson, T. Ericson,
I. Ingemarsson, R. Johannesson,
K. Zigangirov, C.-E. Sundberg
Topics in Coding Theory
VII, 176 pages, 1989

Vol. 129: W. A.Porter, S. C. Kak (Eds.)
Advances in Communications and
Signal Processing
VI, 376 pages, 1989.

Vol. 130: W. A. Porter, S. C. Kak,
J. L. Aravena (Eds.)
Advances in Computing and Control
VI, 367 pages, 1989

Vol. 131: S. M. Joshi
Control of Large Flexible Space Structures
IX, 196 pages, 1989.

Vol. 132: W.-Y. Ng
Interactive Multi-Objective Programming
as a Framework for Computer-Aided Control
System Design
XV, 182 pages, 1989.

Vol. 133: R. P. Leland
Stochastic Models for Laser Propagation
in Atmospheric Turbulence
VII, 145 pages, 1989.

Vol. 134: X. J. Zhang
Auxiliary Signal Design in Fault
Detection and Diagnosis
XII, 213 pages, 1989

Vol. 135: H. Nijmeijer, J. M. Schumacher (Eds.)
Three Decades of Mathematical System Theory
A Collection of Surveys at the Occasion of the
50th Birthday of Jan C. Willems
VI, 562 pages, 1989

Vol. 136: J. Zabczyk (Ed.)
Stochastic Systems and Optimization
Proceedings of the 6th IFIP WG 7.1
Working Conference,
Warsaw, Poland, September 12–16, 1988
VI, 374 pages. 1989

Lecture Notes in Control and Information Sciences

Edited by M. Thoma and A. Wyner